NATURAL DISASTER RESEARCH, PREDICTION AND MITIGATION

FLOODING

RISK FACTORS, ENVIRONMENTAL IMPACTS AND MANAGEMENT STRATEGIES

NATURAL DISASTER RESEARCH, PREDICTION AND MITIGATION

Additional books in this series can be found on Nova's website under the Series tab.

Additional e-books in this series can be found on Nova's website under the e-book tab.

NATURAL DISASTER RESEARCH, PREDICTION AND MITIGATION

FLOODING

RISK FACTORS, ENVIRONMENTAL IMPACTS AND MANAGEMENT STRATEGIES

MOSEKI RONALD MOTSHOLAPHEKO
AND
DONALD LETSHOLO KGATHI
EDITORS

nova publishers
New York

For permission to use material from this book please contact us:
Telephone 631-231-7269; Fax 631-231-8175
Web Site: http://www.novapublishers.com

Library of Congress Cataloging-in-Publication Data

ISBN: 978-1-63117-607-4

Published by Nova Science Publishers, Inc. † New York

CONTENTS

PREFACE

This book is about flooding, the risk it imposes on human well-being and related activities, and the main approaches used to deal with the impacts. The aim is to derive lessons for flood risk management. The book covers experiences from case studies in the five countries of Argentina, Australia, Botswana, Brazil and Taiwan. It indicates that in most areas around the world, floods disrupt human activities and also pose threats to human well-being whereas in other areas, particularly wetlands around the world, they are viewed as useful for the sustainability of ecosystems and human livelihoods. Due to climate variability and change, floods are expected to increase in frequency and intensity throughout the world. There is need to evaluate the current structural and non-structural approaches for dealing with flood risk and the impacts on human systems. Decision-making on the adoption of either structural or non-structural approaches to flood risk largely depends on information available and the means to achieve the intended objectives. Understanding the risk posed by flooding requires multidisciplinary assessments on the biophysical, socioeconomic and cultural factors underlying the vulnerability of human systems. The book starts by identifying some methods which may be useful for flood assessments. Furthermore, it identifies the impacts of flooding and assesses the pros and cons of the related structural and non-structural responses. The challenges observed from the two main approaches are identified and suggestions are made for promoting flood risk management. Suggestions are made for strengthening support for non-structural approaches which are still inadequate in most developing countries, and require improvement in developed countries, given the increasingly complex nature of flood risk posed by extremes in climate variability.

Chapter 1 – Flooding, which occurs when large bodies of water overflow river channels and banks inundating vast areas within and at times beyond the 'normal' river floodplains, is a common phenomenon which occurs in river basins throughout the world. For many centuries to date, floods have been a part of life for many communities and societies which depend on river systems for their development activities and to derive a multiplicity of benefits including water supply, transportation, sewerage disposal, power generation, agricultural production, fishing and so on. As the human population increased, many areas around river systems including floodplains and wetlands have been developed through construction of important infrastructure for public and private use and coversion to agricultural land. Such development has not come without a price; the risk of disruption by floods has been increasing in most areas around the world. Risk itself being the "the probability of an event and its negative consequences" (UNISDR, 2009: p25), is at the interface of the interactions between any event (including floods) and human systems.

Chapter 2 – Flooding is one of the worst natural disasters that cause notable loss of lives and economic damage in Australia similar to many other countries around the world. In Australia, during the 2010-12 periods alone, flood damage costs were estimated at over $20 billion and over 25 human deaths. The flood risk assessment is an essential part of development and operation processes which are subject to flood risk, such as design of hydraulic structures, development control and flood insurance studies. This chapter reviews the commonly adopted flood risk assessment methods adopted in Australia. This, in particular, presents the evolution of a holistic approach of flood estimation known as Joint Probability Approach (JPA)/Monte Carlo Simulation Technique (MCST) in Australia. This method considers the probabilistic nature of flood producing variables and their correlation in flood modelling using a runoff routing model. A case study is presented to illustrate the regionalisation and application of the JPA/MCST to an Australian catchment. It is found that the JPA/MCST can be applied successfully in practice with regional input data. This method can easily be adapted to other countries.

Chapter 3 – The northeast region of the Province of Buenos Aires, Argentina comprises a coastal plain along the Rio de la Plata estuary and an adjacent continental plain. This region exhibits a high degree of urbanization associated with continuously growing economic activities. Regional plain features, wet climate, occurrence of extra-tropical storm surges, and land changes carried out without suitable planning have often produced periodic and disastrous floods in highly urbanized areas particularly those located in topographically low-lying sections. Heavy rainfall events of progressively increasing frequency, the diminishing of infiltration due to extensive urbanization and the subsequent increment in surface runoff have aggravated the problem. This chapter aims at analyzing the causes and environmental impacts of flooding in Province of Buenos Aires. The tragic flooding of the 2nd April 2013 in La Plata has been selected as a paradigmatic case study. Hydro-meteorological monitoring for three locally possible flooding scenarios is considered. A series of structural and non-structural measures that would facilitate the mitigation of flood damage were also analysed. Such measures are recommended for implementation in the continental and the coastal plains.

Chapter 4 – This chapter focuses on the relationship between climate change and the impacts of climate events, as well the importance to manage the associated risk. According to IPCC (2012), the impacts of climate extremes and the potential for disasters emerge from the interaction of physical conditions (weather/climate), from the exposure and vulnerability of human and natural systems. With global climate change, extreme weather events such as cyclones and floods are expected to occur with increasing frequency and greater intensity, which contaminates freshwater supplies, heightens the risk of water-borne diseases, and creates breeding grounds for diseases such as leptospirosis. Infectious disease outbreaks have been reported following major flood events in developing countries, and these outbreaks vary in magnitude and rates of mortality. The objective of the study was to evaluate leptospirosis at geographic locations based on environmental factors and produce a predictive disease risk map. The area of study is São Paulo Metropolitan Area (SPMA), one of the largest megacities in the world (with 19,956,590 people). In this case, authors considered the previous analysis of Coelho and Massad (2012), where for each additional 20 mm of precipitation there was an average increase of 15.6 % in hospital admissions, as the precipitation increased from 20 mm to 120 mm. Within this context, authors considered the spatial distribution of the daily number of leptospirosis cases and the probability of flood events in SPMA. Authors used ArcGIS to integrate the spatial information and non-spatial attribute data, where each spatial feature and its attribute information were linked. The ArcGIS also provided the "math" module to perform mathematical operations in order to analyze the geographical patterns and trends of the region. As a result, a map of vulnerability was produced based on the leptospirosis risk evaluation which can enable decisions to be made on strategies and interventions for improving the conditions of people living in disease-prone areas.

Chapter 5 – Due to dense population living along the riverside, river flooding is one of the major hazards in Taiwan. The Taipei metropolitan area is the largest city with more than one-third of the total population of Taiwan. In this area, numerous severe flooding disasters caused by typhoon events have occurred and resulted in heavy losses. To mitigate flood-related disasters, a large-scale flood prevention program was implemented in 1963 and fully completed in 1999, namely the Taipei Flood Prevention System. The specific goal was to protect the Taipei metropolitan area against the 200-year recurrence flood in the Tanshui River. Levees and dykes of 32 km in length were constructed and improved along the river. Mitigation of floods for the Tanshui River system is inhibited by the bottleneck, which occurs at the smallest river width near the Taipei Bridge. Therefore, the Erchung Floodway was established to divert some of the flood water. The specific goal of the Erchung Floodway was to divert 9200 m^3/sec peak flood discharge under a 200-year return period flood. However Erchung Floodway's function has been changed over time by urban development and natural alterations including river sand mining, riverine park construction, riverine plant succession, bridge construction and so on. Riverbed elevation changes due to sand mining are believed to have influenced the strength of the Taipei Flood Prevention System and thus were discussed in this study. The current protection criteria and impacts from anthropogenic effects and climate change threats were also examined. Both the physical and numerical models were used and analyzed. The sensitivity analysis of thirty-two scenarios corresponding to four factors has been investigated, including riverbed elevation, riverbed roughness, and water stage at the river mouth under the Q_{200} flood. The simulated results show that the flood diversion capacity of the Erchung Floodway, a key infrastructure for dividing floods in the Taipei Flood Prevention System, has decreased by 30%. Authors also found that the Taipei Flood Prevention System will encounter challenges if the riverbed roughness in the Erchung Floodway increases by over 50%, the riverbed roughness in the Tanshui River increases by over 25%, and the Q_{200} increases by over 13%. Authors conclude that the degrading process of the Taipei Flood Prevention System due to rapid urbanization and the corresponding strategies including river roughness and riverbed elevation control are meaningful lessons especially for developing countries. A comprehensive and effective evacuation program and monitoring system is also suggested.

Chapter 6 – In 2011 in Queensland Australia, floods created considerable financial pressure on regional governments. Many affected households suffered severe economic losses as they did not have flood cover on their home insurance policies. The absence of flood insurance could pose threats to fiscal health and has risen to the national policy agenda. This study contributes to the debates by identifying key subjective factors associated with non-insurance. It is based on a social survey involving a total of 501 residents of Brisbane, the Gold Coast, and the Sunshine Coast. A significant minority of respondents (43.8%) reported to have no flood cover. Perceived flood risk is not statistically related to the likelihood of non-insurance. This means that non-insured households are not restricted to those who are unaware of the flood risks confronting them. The insured individuals are better educated and tend to recognize the role of flood insurance in financially protecting the household. Non-insurance is also associated with the expressed preference for government compensation over insurance. The study offers two main insights for policy-makers, floodplain managers and insurers. First, raising risk awareness is unlikely to be sufficient to improving the uptake of flood insurance. Second, managing the public expectations about disaster relief may have positive impacts.

Chapter 7 – Floods and flooding events are of central interest in the studies bordering on the Okavango Delta ecosystems, the sustainability of which depends on regular water flow. Nonetheless, as beneficial as flood pulses might be to the river basin and the riparian communities in and around it, extreme flooding events continue to impact on rural livelihood systems and people's well-being in the area. This chapter employs the concept of Pierre Bourdieu's [1930-2002] habitus and the use of qualitative data (obtained through key informant interviews) to analyse and explain how cultural values shape people's perceptions and how they respond to natural phenomena (such as floods), which impinge on their living conditions. Through the application of Kurt Lewin's [1890-1947] field theory and 3-step model of planned change, and in partial combination with Bourdieu's field, the discourse offers insights on how scheduled change agencies could better understand the social forces that perpetuate undesired and desired behaviours of individuals comprising their clientele systems and how this understanding could enhance the application of appropriate planned change program for achieving behavioural change in the periods of emergency triggered by water inundation.

Chapter 8 – Adaptation to flooding is now widely adopted as an appropriate policy option since flood mitigation measures largely exceed the capability of most developing countries. In wetlands, such as the Okavango Delta, adaptation is more appropriate as these systems serve as natural flood control mechanisms. The Okavango Delta system is subject to annual variability in flooding with extreme floods resulting in adverse impacts on rural livelihoods. This study therefore seeks to improve the general understanding of rural household livelihood adaptation to extreme flooding in the Okavango Delta. Specific objectives are: 1) to assess household access to forms of capital necessary for enhanced capacity to adapt, 2) to assess the impacts of extreme flooding on household livelihoods, and 3) to identify and assess household livelihood responses to extreme flooding. The study uses the sustainable livelihood and the socio-ecological frameworks to analyse the livelihood patterns and resilience to extreme flooding. Results from a survey of 623 households in five villages, key informants, focus group discussions and review of literature, indicate that access to natural capital was generally high, but low for financial, physical, human and social capital. Households mainly relied on farm-based livelihood activities, some non-farm activities, limited rural trade and public transfers. In 2004 and 2009, extreme flooding resulted in livelihood disruptions in the study areas. The main impacts included crop damage, household displacement, destruction of household property, livestock drowning and mud-trapping, the destruction of public infrastructure and disruption of services. The main household coping strategies were labour switching to other livelihood activities, temporary relocation to less affected areas, use of canoes for early harvesting or evacuation and government assistance, particularly for the most vulnerable households. Household adaptive strategies included livelihood diversification, long-term mobility and training in non-agricultural skills. The study concludes that household capacity to adapt to extreme flooding in the study villages largely depends on access to natural capital. This is threatened by population growth, land use changes, policy shifts, upstream developments, global economic changes and flood variations due to climate variability and change.

Chapter 9 – Flooding is a global phenomenon that continues to affect people socially, psychologically and economically and has been a concern for many years. Risk communication has been applied to inform the public about the potential harm caused by different environmental hazards. Literature has suggested that in cases where potential harm is perceived, the effectiveness of the risk communication interventions will depend on a number of factors such as perception of risk, trust, credibility and socio-demographic and cultural factors. The current study aimed at exploring factors that contribute to low adoption of flood risk warning information among local communities living in the Okavango Delta. The qualitative study, using a mixed method approach, was guided by the Mental Models and Trust Determination model to explore the socio-cultural and institutional environment within which flood risk communication took place, using a sample of 55 respondents affected by the floods. Findings revealed multiple factors such as a history of long residence in the area without experiencing floods, socio-cultural, myths, beliefs and perceptions towards water and floods, and low knowledge about floods and floods risks as factors contributing to the low adoption of flood risk warning information among the Delta communities. Others emanating from the sources of risk communitors include issues of trust and credibility. The study suggests that risk communicators and institutions should undertake preliminary synoptic audience assessments to guide communication interventions, participatory communication and land suitability and risk vulnerability before allocations.

Chapter 10 – Fish is a major source of livelihoods for riparian communities. It contributes significantly to food and nutrition security, rural employment and general poverty alleviation. Studying the dynamics of this resource in riverine/flood pulsed systems provides critical information essential for management of this resource. It is also important to understand the impact of extreme flood events on this resource, so that their impact on the rural livelihoods can be established. In the absence of comprehensive data linking the effects of extreme floods to riverine fish dynamics, predictions can be made using holistic environmental flow assessment (EFA) methodologies. Therefore, this approach was used to assess the effect of extreme floods on the Boteti River fishery. Typical of holistic methods in EFA methodologies, baseline information on the biology and ecology of Boteti River fisheries were determined. Fish data were collected using standardised experimental fishing methods, while socio-economic data were collected using structured interviews. Various fish community indices were used to assess spatio-temporal variations in fish community structure and feeding ecology of selected species. Two fish indicators, based on fish guilds, were then used to predict the effect of extreme flooding on fish populations. Qualitatively, results revealed that there were spatio-temporal variations in fish community structure in the river systems, though there were no significant differences ($p>0.05$) in selected indices among the different sites. However, there were significant differences ($p<0.05$) in relative abundances and morphometrics of selected species among sites in the river system. This suggests that variability in the fish community is observed at the species level and not the population level. Results also revealed that the terrestrial environment is a major source of energy for the river's fish community, driven by flooding. Understandably, the fish resource in this river is a major source of livelihoods for the riparian community, though fishers had diverse economic activities. It was predicted that while extreme floods had negative impacts on fish populations, prolonged flooding also contributes to increased fish production for some species. While negative impacts on fish production would have a deleterious impact on fishers, it was also concluded that their diverse economic activities is an adaptation strategy to variability in fish availability. Another major conclusion is that the Boteti River fishery is very resilient (and dynamic), and will always return to normal after any extreme event such as drought or flood.

Chapter 11 – This book highlights the international experience of flood risk in five countries of Argentina, Australia, Botswana, Brazil, and Taiwan. It demonstrates that flooding is one of the worst hazards in both the developed and developing world. Floods are rated the third most common natural disaster after storms and earthquakes. Despite the global efforts to manage floods and mitigate their damages, there is evidence that the losses caused by floods continue to increase. The book indicates that flooding is not only caused by too much water but results from a number of conditions. These include extreme meteorological events and poor land-use planning (Chapter 2). Based on the predictions of IPCC (2012), extreme weather events are expected to increase in the future as a result of climate change and this implies that they are likely to exacerbate the damages resulting from floods. Apart from these challenges, floods have also created opportunities as they enrich the land for agriculture by depositing nutrient-rich sediments in floodplains and therefore contribute to the provision of ecosystem services. Since the 1980s, many countries realised need to live in harmony with the processes of flooding as it is almost impossible to eliminate them. The book supports the view of Sayers et al. (2013) that there is no single solution for addressing floods and that it is necessary to develop a portfolio of measures. Risk-based approaches are recommended as a strategy for addressing the challenges of flooding.

LIST OF ACRONYMS

ABS	Australian Bureau of Statistics
A-D	Anderson-Darling test
ADCP	Acoustic Doppler Current Profiler
ADRC	Applied Development Research Consultants
AHDEL	American Heritage Dictionary of the English Language
ARI	Average Recurrence Interval
ARR	Australian Rainfall and Runoff
AUD$	Australian Dollar
BOM	Australian Bureau of Meteorology
BWP	Botswana Pula
CATI	Computer Assisted Telephone Interviewing
CBNRM	Community-based natural resource management
CEU	Cattle-equivalent-units
CRCCH	Cooperative Research Centre for Catchment Hydrology
CRED	Centre for Research on the Epidemiology and Disasters
C-S	Chi-Square test
CSO	Central Statistics Office
DDMC	District Disaster Management Committee
DEA	Design Event Approach
DSS	Decision support system
DFFC	Derived flood frequency curve
DOD	District Officer Development
EA	Enumeration area
EFA	Environmental flow assessment
EWP	Electronic White Pages
FAO	United Nations Food and Agricultural Organisation
FFA	Flood frequency analysis
FGD	Focus Group discussion
GDP	Gross Domestic Product

GIS	Geographic Information Systems
HDI	Human development index
HIV/AIDS	Human immunity virus/Acquired immunity deficiency syndrome
ICA	Insurance Council of Australia
IED	Inter-event duration
IFD	Intensity-frequency-duration
IFRC	International Federation of Red Cross
IPCC	Intergovernmental Panel on Climate Change
JPA	Joint Probability Approach
IRI	Index of relative importance
K-S	Kolmogorov-Smirnov test
LIPW	Labour-intensive-public-works
MCST	Monte Carlo Simulation Technique
MSL	Mean sea level
NCWE	National Committee on Water Engineering (Australia)
NDIR	Natural Disaster Insurance Review (Australia)
NGOs	Non-governmental organizations
N-NW	North-northwest
NSW	New South Wales
RDD	Random Digit Dialling
RORB	Australian Rainfall Runoff Routing model
SADC	Southern African Development Community
SE-SSE	Southeast south-south-east
SHN	Servicio de Hidrografia Naval
SNWA	Seven Natural Wonders of Africa
SPF	Strategic Programme Fund
SPSS	Statistical Package for Social Sciences (software)
SPMA	São Paulo Metropolitan Area
TAR	Third Assessment Report
TP	Temporal patterns
TPT	Total Probability Theorem
UK	United Kingdom
UNDP	United Nations Development Programme
UNISDR	United Nations International Strategy for Disaster Reduction
UNESCO	United Nations Education Scientific and Cultural Organisation
URBS	Unified River Basin Simulator
US$	United States Dollar
VDC	Village development committee
WMO	World Meteorological Organisation
ZST	Zambezi Safari Travels

LIST OF TABLES

LIST OF FIGURES

In: Flooding
Editors: M. R. Motsholapheko and D. L. Kgathi
ISBN: 978-1-63117-607-4

Chapter 1

FLOODING, RISK FACTORS AND RESPONSES: AN OVERVIEW OF CONCEPTS

M. R. Motsholapheko ***and D. L. Kgathi***
Okavango Research Institute, University of Botswana

Flooding, which occurs when large bodies of water overflow river channels and banks inundating vast areas within and at times beyond the 'normal' river floodplains (Schanze, 2006), is a common phenomenon which occurs in river basins throughout the world. For many centuries to date, floods have been a part of life for many communities and societies which depend on river systems for their development activities and to derive a multiplicity of benefits including water supply, transportation, sewerage disposal, power generation, agricultural production, fishing and so on (Sayers et al., 2013). As the human population increased, many areas around river systems including floodplains and wetlands have been developed through construction of important infrastructure for public and private use and coversion to agricultural land (Sayers et al., 2013; Hartig et al., 1997).Such development has not come without a price; the risk of disruption by floods has been increasing in most areas around the world (Sayers et al., 2013; Wisner et al., 2004). Risk itself being the "the probability of an event and its negative consequences" (UNISDR, 2009: p25), is at the interface of the interactions between any event (including floods) and human systems.

In current times more than any other time in the past, the level of risk and impacts that flooding imposes on human well-being and related activities of economic development may be what makes it a blessing on one point of a continuum or a disaster on the other end of the same continuum. In some areas, particularly wetlands around the world, floods are viewed as useful for the sustainability of ecosystems and human livelihoods (Wisner et al., 2004; Samuels et al., 2006; Hartig et al., 1997). Throughout the world floods are also known to destroy human life and cause disruptions to human activities more than most natural hazards, being third after storms and earthquakes (World Bank/United Nations, 2010). Floods have in past and recent decades killed thousands of people and continue to be the main danger for human life (World Bank/United Nations, 2010; Jonkman, 2005; Wisner et al., 2004). This is the case particularly in developing countries, where there is general lack of capacity to

* Corresponding author address: P/Bag 285 Maun, Botswana. E-mail: rmoseki@ori.ub.bw

mitigate or adapt to the impacts. Due to climate variability and change, floods are expected to increase in frequency and intensity throughout the world and the impacts of flooding may be felt in both wet and semi-arid regions. According to the IPCC (2007) the possibility of vicious flash floods exists in semi-arid countries, due to increased severity of storms. The risks of flooding may increase with the resultant increase in the related losses in the form of damaged infrastructure and household property, human deaths and so on.

The risk factors of flooding can be distinguished into two main types. These include the factors which relate to the biophysical nature of flooding as a hydrological event, and those which are associated with human systems' vulnerability to the impacts of flooding. These two forms of risk factors fit well within the IPCC (2001: p95) definition of vulnerability as ''a function of the character, magnitude, and rate of climate variation to which a system is exposed, its sensitivity, and its adaptive capacity''. The definition points to both the biophysical nature of climatic variations as well as the susceptibility to the impacts. Climatic variations, particularly extremes are common precursors for natural hazards such as severe storms, hurricanes and others which may result in hydrological events such as floods. Biophysical risk factors for events such as floods are largely predictable through various methods and/or early warning systems, whereas the human responses are amassed with uncertainties. This is because human responses to flooding and other forms of risk are generally complex, as they may have physiological, psychological, religious, cultural and socio-economic dimensions, which may in turn have spatial and temporal differences.

It is these differences and the realisation that floods pose increasing risks to human well-being and related activities, that has led to continuous change in the strategies for mitigation and adaptation to the impacts of flooding. Past approaches to flood risk involved direct control of water flow and protection against floods through construction of levees, dykes and dams, channel diversions and other engineering works (Sayers et al., 2013). In the middle of the twentieth century, there was a major paradigm shift from flood protection to flood risk management (Johnson and Priest, 2008; Sayers et al., 2013). The advent of major drivers such as climate change and the development of the concept of sustainable development in the 1980s contributed to this paradigm shift. Climate change contributed to this shift as one of the factors which exacerbates flooding through increased climate variability and frequency of extreme events. Sustainable development contributed to the shift in paradigm through its emphasis on the reduction of the impacts, of economic development and other human activities, on the environment. It addition, sustainable development promotes the wise use of floodplains which are a major source of ecosystem services.

The adoption of flood risk management through major changes in decision-making helped many countries and regional water governing organisations to shift from flood control measures solely based on "hard engineering" (structural approaches) to those which had a mixture of both structural and non-structural measures. The former approach had limitations in that it emphasised control rather than management; it was undertaken in isolation and on *ad hoc* basis, as well as being reactive rather than proactive, and it developed solutions which were largely monodisciplinary (United Nations Economic Commission for Europe [UNECE, 2009), The latter approach considered environmental impacts and also promoted policies on flood warning, mapping of the probability of flooding, flood insurance for those at risk of floods and wise use of floodplains through appropriate planning (Johnson and Priest, 2008; Sayers et al. 2013). In the 1990s the concept of risk management gained more importance and started to be used in flood management, worldwide. For example in the EU, a Habitats

Directive which addressed the importance of the environmental impacts in flood management came into force in 1992 (Sayers et al., 2013).The European Commission issued a Flood Risk Directive in October, 2007 in order to address the problem of flooding in the member states (Mostert and Junier, 2009). The Directive required member states to undertake preliminary flood risk assessments of their river basins, in the first stage, in order to determine potential areas of flood risk. Member states were also required to prepare flood hazard and flood risk maps in the second stage, and flood risk management plans in the third stage (Mostert and Junier, 2009). China, a country with two thirds of the land prone to floods, also adopted a strategy of managing floods rather than trying to prevent them. The awareness of people about floods was promoted by the mapping of flood hazard areas (Sayers et al., 2013).

The concept of flood risk management is evolving and it considers both the nature of the flood event and the consequences emanating from the impacts (Schanze, 2006; Meyer et al., 2009). According to Schanze et al. (2006), three tasks could be done to structure the activities of flood risk management: flood risk analysis, flood risk assessment and flood risk reduction. Flood risk analysis examines the past, present and future aspects of flood risk using both natural and social science methods. The current known methods of risk analysis include statistical analysis of hydro-meteorological events, economic analysis of the losses resulting from floods, mapping of flood risk areas and forescasting and developing early warning systems (Schanze et al. 2013). Flood risk assessment determines the perceptions and attitudes of those who are at risk of flooding. The assessment is based on socio-cultural and psychological-cognitive perceptions of those who are affected by flood risk. Research on flood risk communication is one of the key aspects of flood risk assessment. Flood risk reduction entails all actions aimed at minimising the impacts of floods, either in anticipation or in response to flooding (Schanze, 2006). Measures of flood risk reduction include structural methods such as the construction of levees, diversions and sluices, and non-structural measures such as land-use planning, adaptive management and early warning systems (Wilby and Keenan, 2012).

A further improvement in the management of flood risk was development of integrated flood management which recognises that, although floods pose risk to human life and related activities, they are crucial for ecosystem service provision, they cannot be completely be curbed and that they are an integral part of the hydrological cycle, which needs to be managed through an ecosystem approach (UNECE, 2009; World Meteorological Organisation [WMO], 2009). This new thinking was influenced by the concept of integrated water resources management which promotes coordinated rather fragmented efforts to manage water, land and related resources within a river basin as the main ecological unit (GWP-TAC, 2000). This approach also recognises that floods are not confined by international boundries and that management efforts upstream may have impacts downstream, hence the development of transboundary flood management which is being promoted by WMO in partnership with Global Water Partnership (GWP) in many regions around the world (WMO, 2009).

This book serves as a compendium of case studies from both developed and developing countries with the aim to derive lessons for flood risk management now and in the future. It is arranged into chapters which seek to improve understanding of flooding and the nature of anticipative and reactive responses to floods. It is hoped that through understanding the varying conditions within which flooding occurs in countries around the world, appropriate policies can be developed towards combating the risk and minimizing its impacts.

The book is arranged as follows. Chapter 1 introduces the key concepts and attempts to lay the approach to the study of flooding as a natural phenomenon and a shock to human systems' within the global context. Chapter 2 presents the Joint Probability Approach/Monte Carlo Simulation Technique (JPA/MCST) a method of flood assessment developed for use in in Australia and has potential to be applied in various parts of the world. Chapter 3 illustrates how human activities influenced the natural water flows and magnified the impacts of flooding in the north-east region of the Buenos Aires Province of Argentina. Chapter 4 assesses the impacts of floods and provides a mapping of areas vulnerable to leptospirosis which is a disease associated with flooding in the Sao Paulo metropolitan area of Brazil. Chapter 5 assesses the human-induced changes to river flows and the threats to the flood defence systems in Taipei, Taiwan.

Chapter 6 assesses the influence of risk perceptions on household decision-making on the adoption of flood insurance in Australia. Chapter 7 uses the concept of habitus to assess the influence of culture on the adoption of flood risk early warning messages in the Okavango Delta Botswana. Through Kurt Lewins' field theory and three-step model of planned change, some suggestions are made for a shift in approach to flood risk management that recongnises the cultural, psychological, social, economic and other conditions that are peculiar to the areas where it is undertaken. Chapter 8 assesses the impacts of, and household level adaptation to, extreme flooding in the Okavango Delta of Botswana. Chapter 9 uses the Mental Models and Trust Determination model to determine factors that influence adoption of flood risk information in flood-prone rural communities of the Okavango Delta, Botswana. Chapter 10 assesses the dynamics of the Okavango Delta fishery. It also determines the impact of extreme flooding on fish resources, and the subsequent impacts on rural livelihoods in the Boteti River, an outflow channel of the Okavango Delta in Botswana. Chapter 11 concludes and suggests implications for flood risk response within regional and global contexts.

REFERENCES

Global Water Partnership Technical Advisory Committee (GWP-TAC), (2000) Integrated water resources management. *Technical background paper number 4*. Stockholm: Global Water Partnership.

Hartig, E. K., Grozev, O., Rosenzweig, C., (1997). Climate change, agriculture and wetlands in Easter Europe: vulnerability, adaptation and policy. *Climatic Change*, 36: 107-121.

Inter-governmental Panel on Climate Change [IPCC], (2007). *Climate Change 2007: Impacts, adaptation and vulnerability.* Contribution of Working Group II to the Fourth Assessment Report of the Intergovernmental Panel on Climate Change. Cambridge, UK: Cambridge University Press.

Inter-governmental Panel on Climate Change [IPCC], (2001). Climate change 2001: impacts, adaptation and vulnerability. Summary for policy makers. <http://www.ipcc.ch/pdf /climate-changes-2001/impact-adaptation-vulnerability/impact-spm-en.pdf> [Accessed 12 December 2013].

Johnson, C.L., Priest, S.J., (2008). Flood risk management in England: a changing landscape of risk responsibility? *International Journal of Water Resources Development*, 24(4): 513-525.

Jonkman, S., (2005). Global perspectives on loss of human life caused by floods. *Natural Hazards,* 34:151–75.

Meyer, V., Haase, D., Scheuer, S., (2009). Flood risk assessment in European river basins-concept, methods and challenges exemplified at the Mulde River. *Integrated Environmental Assessment and Management*, 5(1): 17-26.

Mostert, E., Junier, S.J., (2009). The European flood risk directive: challenges for research. *Hydrology and Earth system Sciences discussions*, 6: 4961-4988.

Samuels, P., Klijn, F., Dukman, J., (2006). An analysis of the current practice of policies on river flood risk management in different countries. *Irrigation and Drainage*, 55: 5141-5150.

Sayers, P. Yuanyuan, L., Galloway, G., Fuxin, S., Kang, W., Yiwei, C., Quense, T.L., (2013). *Flood Risk Management: a strategic approach*. Paris: UNESCO.

Schanze, J., (2006). Flood risk management- a basic framework. In: Schanze, J., Zeman, E., Marsalek, J. (Eds) *Flood risk management- hazards, vulnerability and mitigation measures* (pp1-20). Berlin, Germany: Springer.

Schanze, J., Zeman, E., Marsalek, J. (Eds), (2006). *Flood risk management- hazards, vulnerability and mitigation measures*. Berlin, Germany: Springer.

United Nations International Strategy for Disaster Risk Reduction [UNISDR], (2009). UNISDR terminology on disaster risk reduction. Geneva: UNISDR.

United Nations Economic Commission for Europe [UNECE], (2009). Transboundary flood risk management: experiences from the UNECE region. New York and Geneva: UNECE.

Wilby R.L., Keenan, R., (2012). *Adapting to flood risk under climate change. Progress in Physical Geography*, 36(3): 348-378.

Wisner, B., Blaikie, P., Cannon, T., Davis, I., (2004). At *risk: natural hazards, people's vulnerability and disasters* (2nd Edition). London: Routledge.

World Bank/United Nations, (2010). *Natural unnatural disasters, the economics of effective prevention.* Washington DC: World Bank.

World Meteorological Organisation [WMO], (2009). *Integrated flood management concept paper.* Geneva, Switzerland: World Meteorological Organisation.

In: Flooding
Editors: M. R. Motsholapheko and D. L. Kgathi
ISBN: 978-1-63117-607-4

Chapter 2

FLOOD RISK ASSESSMENT IN AUSTRALIA: APPLICATION OF A HOLISTIC APPROACH

W. L. Caballero, M. M. Haque and A. Rahman*
University of Western Sydney, Australia

ABSTRACT

Flooding is one of the worst natural disasters that cause notable loss of lives and economic damage in Australia similar to many other countries around the world. In Australia, during the 2010-12 periods alone, flood damage costs were estimated at over $20 billion and over 25 human deaths. The flood risk assessment is an essential part of development and operation processes which are subject to flood risk, such as design of hydraulic structures, development control and flood insurance studies. This chapter reviews the commonly adopted flood risk assessment methods adopted in Australia. This, in particular, presents the evolution of a holistic approach of flood estimation known as Joint Probability Approach (JPA)/Monte Carlo Simulation Technique (MCST) in Australia. This method considers the probabilistic nature of flood producing variables and their correlation in flood modelling using a runoff routing model. A case study is presented to illustrate the regionalisation and application of the JPA/MCST to an Australian catchment. It is found that the JPA/MCST can be applied successfully in practice with regional input data. This method can easily be adapted to other countries.

Keywords: flood, joint probability approach, Monte Carlo simulation, regionalization

INTRODUCTION

Flooding is one of the worst natural disasters causing notable loss of lives, economic damage and disruption to society. The damages caused by floods are substantial on both local

* School of Computing, Engineering and Mathematics, University of Western Sydney, Building XB, Kingswood, Locked Bag 1797, Penrith, NSW 2751, Australia.
Email: a.rahman@uws.edu.au.

and global scale. Some of the recent flooding events in different countries in the world have shown disastrous effect on human lives and properties as well as on the overall economy of some countries. Hurricane Katrina and Cyclone Sidr caused catastrophic flooding in South East Louisiana in USA and coastal regions of Bangladesh, respectively, killing more than a thousand people and leading to notable economic losses (Jonkman and Vrijling, 2008).

In 2011 floods were considered as the third most common disasters after earthquakes and tsunamis (Centre for Research on the Epidemiology and Disasters [CRED], 2012). In Australia, floods are considered to be the among the most common and devastating natural disasters (FitzGerald et al., 2010). In the period 1852 to 2011, 951 people were killed and another 1326 injured by floods in Australia (Carbone and Hanson, 2013). Average costs of flood damage in Australia was estimated to be around $400 million per annum (Charalambous et al., 2013). Floods do not only have adverse effects on local communities but also impact on the overall economy of a country. For example, in Queensland, about 70% of the state was impacted on by floods and around 75% of the banana crop was destroyed in the period 2010 to 2011. Total damage to the infrastructure was estimated to be around AUS$5 billion (Price Waterhouse Cooper , 2011). Therefore, development of a reliable flood risk assessment technique is crucial to reduce the flood damage cost and fatalities.

Flood risk assessment is needed in many development applications. It can help to manage flood risk, design mitigation plans and emergency strategies. In Australia, the Design Event Approach (DEA) which is a rainfall based flood estimation method is currently recommended by the Australian Rainfall and Runoff (ARR) to estimate complete design flood hydrograph (Institution of Engineers Australia, 1987). This approach attempts to convert a design rainfall event to a flood hydrograph event using a runoff routing type hydrological model. The DEA has serious shortcomings (Weinmann et al., 1998; Rahman et al., 2002a, 2002b; Nathan et al., 2003; Kuczera et al., 2006; Svensson et al., 2013). The key limitations of the DEA stems from the fact that design rainfall input for a given Average Recurrence Interval (ARI) can be transformed into a design flood output of the same ARI. This can be done using the representative values of design inputs such as rainfall temporal pattern and loss parameters. It implies that these input variables are probability neutral. However, it has been found that rainfall temporal patterns and losses can show a wide variability over space and time and these need to be considered as stochastic variables in modelling.

To overcome these limitations with the DEA and to make improvements in design flood estimates, a more holistic approach based on the Joint Probability Approach (JPA) has been developed in the last decade in Australia (e.g., Rahman et al., 2002a; Mirfenderesk et al., 2013). The fundamental principle of JPA is that it considers the stochastic nature of the flood producing variables and their joint probabilities to estimate probability-distributed flood output (Caballero and Rahman, 2013). The Monte Carlo Simulation Technique (MCST) is a simplified form of the JPA in that it produces the distribution of flood outputs by simulating many possible rainfall-runoff events that are likely to produce floods on a given catchment (Loukas et al., 1996).

A number of research studies have been undertaken on the JPA to develop and facilitate its application in design flood estimation problems (Rahman et al., 2002a; Aronica and Candela, 2007; Kjeldsen et al., 2010). Rahman et al. (2002a) applied a simplified MCST based on the principles of JPA with a runoff routing model to the three catchments in Victoria. They found that this method produced relatively precise flood frequency curves and can overcome some of the shortcomings associated with the DEA. Several studies (Muzik,

2002; Weinmann et al., 2002; Svensson et al., 2013; Charalambous et al., 2013; Caballero and Rahman, 2013) adopted MCST in flood modelling. The JPA/MCST has also been investigated by a number of international studies (Gioia et al., 2008; Haberlandt et al., 2008; Viglione and Bloschl, 2009; Iacobellis et al., 2011). Based on these studies, it has been found that JPA produced more precise and realistic flood estimates than the traditional event based approach. The consideration of the stochastic nature of the flood producing input variables in design flood estimation in Australia was recognised by the ARR (Institution of Engineers Australia, 1987). It recommended the DEA for design flood estimation in 1987. Many studies (Hill and Mein, 1996; Loveridge and Rahman, 2012; Kuczera et al., 2006; Caballero et al., 2011; Caballero and Rahman, 2013; Charalambous et al., 2013) have pointed out the importance of adopting a JPA in design flood estimation. Moreover, the National Committee on Water Engineering in Australia has recently resolved that MCST should be adopted in Australia in preference to the DEA type rainfall-based design flood estimation methods (National Committee on Water Engineering [NCWE], 2013).

The major advantage of the MCST is that it considers the stochastic nature of all the principal flood producing variables. This makes it possible to study the impacts of numerous combinations of input variables in the hydrologic modelling whereas the DEA considers only one design event and ignores the stochastic nature of other input variables except rainfall depth.

Two main approaches are currently being investigated in Australia on the application of JPA: (i) application of the Total Probability Theorem (TPT), called TPT-MCST; and (ii) the approach proposed by the Cooperative Research Centre for Catchment Hydrology (CRCCH), called CRCCH-MCST.

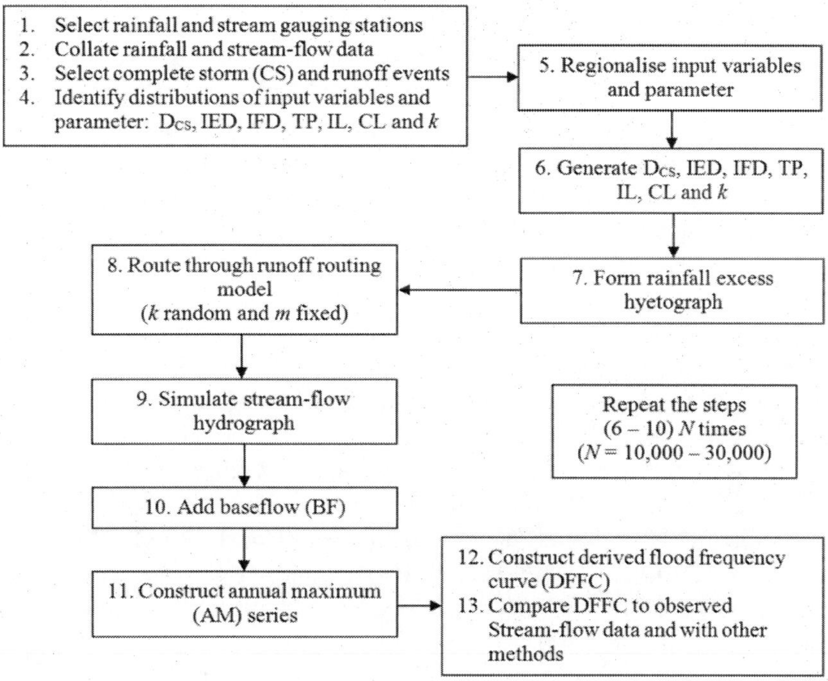

Figure 2.1. Schematic diagram of the adopted MCST methodology.

The TPT-MCST was introduced by Nathan, Weinmann and Kuczera (Laurenson et al., 2005) whereas the CRCCH-MCST approach was developed by Rahman et al. (2002a). In the CRCCH-MCST, thousands of rainfall events are simulated from the marginal distributions of rainfall duration, depth and temporal patterns and losses, which are then fed into a calibrated runoff routing model. The generated flood peaks are then used to construct a derived flood frequency curve. This method is illustrated in Figure 2.1. In the other approaches (i.e., TPT-MCST), flood frequency curves are obtained by simulating random values of temporal pattern and losses while using the fixed value of rainfall duration and intensity data (e.g., Sih et al., 2012; Mirfenderesk et al., 2013; Nathan and Weinmann, 2013). Though TPT-MCST approach does not consider the stochastic nature of all the input variables; it is useful in practice as it uses the widely available intensity-frequency-duration (IFD) data, and is less data hungry than the CRCCH-MCST.

This chapter reviews the commonly adopted flood risk assessment methods in Australia. In particular, it presents the evolution of an holistic approach to design flood estimation known as JPA/MCST in Australia. A case study is presented that illustrates the application of the JPA/MCST to an Australian catchment.

Design Event Approach

The ARR1987 recommended Design Event Approach (DEA) for design flood hydrograph estimation to be adopted in Australia (Institution of Engineers Australia, 1987). This approach uses a number of trial rainfall durations and their corresponding average rainfall intensity values along with fixed values of other input variables (e.g., temporal pattern and losses) to obtain design flood estimates. Design rainfall for a number of durations was routed through a rainfall-runoff model to estimate peak flows for a specified location and ARI. The largest peak flow rate producing rainfall duration is considered as the 'critical duration'. The peak flow associated with this critical duration is taken as the design flood of the specified ARI. The fixed values of the input variables are selected in such a way that the resulting flood peak would have the same ARI as the corresponding design rainfall intensity input. No specific guidelines have been so far established to select the single value of an input variable as it generally shows a wide variability from event to event. The choice of the representative value of a variable is dependent on the preference of the individual designer and it is subject to various assumptions. The inappropriate choice of an input variable is likely to result in a highly biased design flood estimate.

In brief, the DEA only considers the stochastic nature of rainfall depth in design flood estimation, and overlooks the stochastic nature of other input variables. Due to these limitations, the DEA has been criticised by many studies (Weinmann et al., 1998; Rahman et al., 2002a; Kuczera et al., 2006). In Australia, several rainfall-runoff models such as RORB (Laurenson et al., 2005), Unified River Basin Simulator [URBS] (Carroll, 2007) and Watershed Bounded Network Model [WBNM] (Boyd et al., 1987) are used in conjunction with the DEA to estimate the design floods. The former two models have already incorporated the MCST in design flood estimation to some degree.

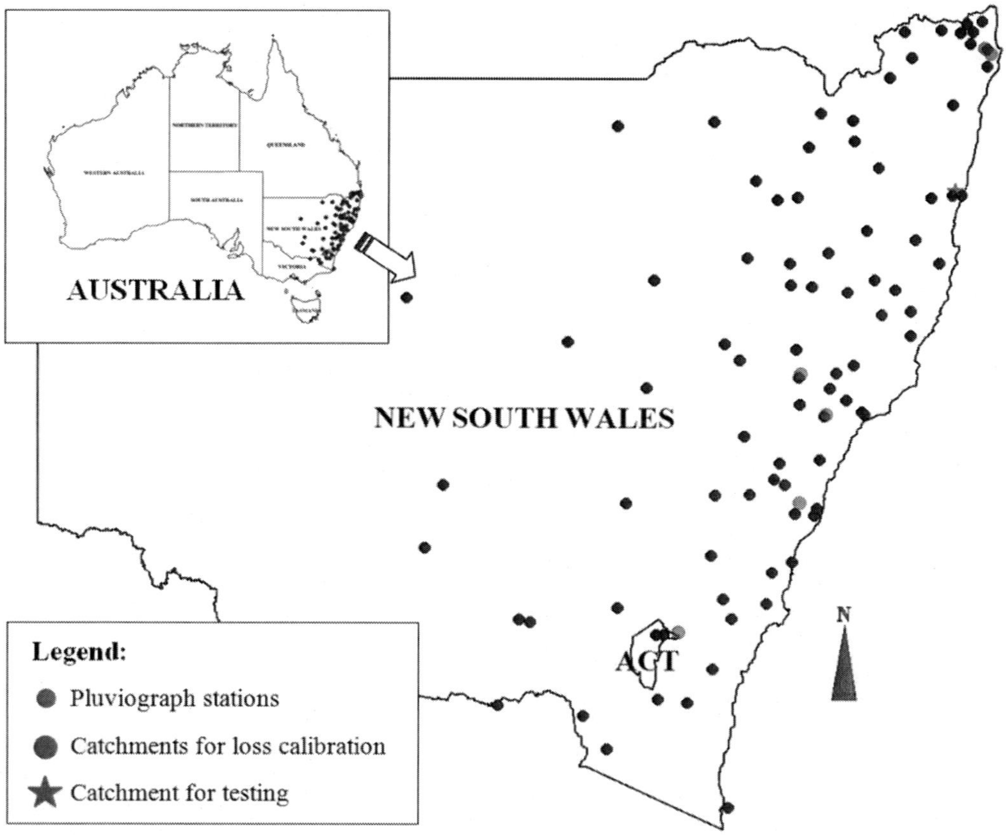

Figure 2-2. Selected 86 pluviograph stations and seven study catchments in New South Wales, Australia.

A Case Study in Australia

This chapter presents a case study using data from the state of New South Wales, Australia. A total of 86 pluviograph stations from New South Wales were selected to regionalise stochastic rainfall duration, intensity and temporal patterns (Figure 2-2). To regionalise losses and runoff routing model parameter, six catchments were selected (as listed in Table 2-1 and shown in Figure 2-1). Furthermore, the Orara River catchment (bold text in Table 2-1) is used to test the applicability of the regionalised stochastic inputs/parameters in the JPA/MCST. The selected catchments have relatively longer records of stream-flow (at least 26 years), which is necessary for accurate at-site flood frequency analysis (FFA). From Table 2-1, it can be seen that the stream-flow data record length ranges from 26 to 46 years with an average of 35 years. The selected catchments have concurrent record length (for pluviograph and stream-flow data) of more than 20 years as indicated in this table.

To develop stochastic rainfall input variables, continuous reading of rainfall data is necessary, and hence 86 pluviograph stations were selected. These data were obtained from Australian Bureau of Meteorology (BOM).

Table 2.1. Selected catchments in New South Wales, Australia

Catchment station ID	Station name	Area (km^2)	Streamflow records	Pluviograph station ID	Pluviograph records	Concurrent record length (years)
203002	Coopers Creek at Repentance	62	1976 - 2009	58072	1965 - 1998	22
203012	Byron Creek at Binna Burra	39	1977 - 2009	58072	1966 - 1998	21
204025	Orara River at Karangi	135	1970 - 2011	59026	1970 - 2010	40
210068	Pokolbin Creek at Pokolbin Site 3	25	1963 - 2009	61238	1962 - 2011	46
210076	Antiene Creek at Liddell	13	1968 - 2010	61212	1964 - 1995	27
213005	Toongabbie Creek at Brien Road	70	1979 - 2009	67035	1965 - 2001	22
411001	Mill Post Creek at Bungendore	16	1959 - 1985	70014	1937 - 2010	26

The pluviograph stations were selected based on a minimum threshold record length of 30 years. This threshold record length was selected to get a good number of events in a given station so that a meaningful statistical test can be used to determine the probability distribution of the rainfall characteristics at a selected station. The rainfall record lengths of the selected 86 pluviograph stations range from 30 years to 101 years with an average record length of 45 years. In Figure 2-2, the selected pluviograph stations present a very good spatial distribution over the eastern part of the state. However, there is no station selected from the far western part of New South Wales as there is no pluviograph station with sufficient time series data of good quality in the region.

SELECTION OF STORM EVENTS

In this chapter, complete storm is considered in the application of JPA/MCST as described in Caballero and Rahman (2013). Complete storm is defined as the significant period of rain preceded and followed by an arbitrary selected period of dry hours for example 6 hours (Rahman et al., 2002a). The inter-event duration is defined as the period (time elapsed in hours) between the successive selected complete storm events. In this analysis, a rainfall intensity threshold is adopted to select an appropriate number of complete storm events, in which it is 0.40 times of the 2-year ARI design rainfall intensity of the corresponding rainfall duration. Using the adopted threshold value, the selection results in an average of two to eight complete storm events per year at a given pluviograph station.The selected complete storms are then analysed to derive the database of rainfall complete storm duration (D_{CS}), inter-event duration (IED), intensity-frequency duration (IFD) and temporal patterns (TP). These data form the basis of stochastic rainfall input variables in the application of JPA/MCST.

Selection of Loss and Runoff Routing Model Parameter

This study adopts the initial loss (IL) and continuing loss (CL) model to generate rainfall excess hyetograph. In this model, the IL is defined as the amount of rainfall that occurs before the start of the runoff, while CL is defined as the average rate of loss throughout the remainder of the rainfall event. This adopts a runoff routing model with a storage delay parameter (k), as given below:

$$S = kQ^m \tag{1}$$

where S is the storage (m^3), k is the storage delay parameter in hours (h), Q is the discharge (m^3/s) and m is the non-linearity parameter, which is taken as 0.8 in this application. Here, the adopted model considers a single storage concentrated at the catchment outlet, hence this method is only applicable to small catchments in the order of 200 km^2. This model is calibrated based on the selected concurrent rainfall and streamflow events. The IL, CL and k values obtained from the calibrated events are used to derive stochastic input database of these variables in the application of the JPA/MCST. This study treats seven random input variables in the application of JPA/MCST (unlike with the DEA which has only one random input, the rainfall depth). These input variables are the D_{CS}, IED, IFD, TP, IL, CL and k.

Regionalisation of Model Inputs and Parameter

The derived input values from the selected rainfall and streamflow events are examined to fit an appropriate probability distribution to each of the seven input variables/model parameters. In this study, exponential and gamma distributions are adopted as candidates to describe the observed data of various input variables similar to Caroll and Rahman (2004) and Haddad and Rahman (2011). The goodness of fit of a hypothesised distribution is tested by applying three goodness-of-fit tests: Chi-Square (C-S) test, Kolmogorov-Smirnov (K-S) test and Anderson-Darling (A-D) test. These tests are applied at the 5% level of significance to examine the statistical hypothesis that the data follows a hypothesised probability distribution. The non-dimensional TP data (similar to Rahman et al., 2002a) are adopted to create a regional TP database that is used to select a random TP during simulation. For all the selected complete storms, the TP data from the selected pluviograph stations are pooled together to create a regional TP database. Here, the TP are expressed in 10 time intervals and cumulative proportion of rainfall depth similar to Rahman et al. (2002a).

The practical application of JPA/MCST needs regionalisation of the model inputs and parameters so that at any arbitrary location in the study region, these model inputs/parameters can be estimated without the need of direct data analysis. This study adopted a spatial proximity method in the regionalisation following an approach similar to Merz and Bloschl (2004), Vaze et al. (2011) and Haddad and Rahman (2012). The D_{CS}, IED and IFD data were regionalised by applying the inverse distance weighted averaging method using observed data from an arbitrary number of nearby stations. Here, the D_{CS}, IED and IFD data of the surrounding pluviograph stations (within or near the catchments' centre, within 30 km radius) were considered in the weighted averaging. Distances greater than 30 km were not considered in the analysis. The regional TP data were pooled using TP data of pluviograph stations within 200 km or a maximum of 20 nearby pluviograph stations. The IL, CL and *k* data are regionalised using the respective distributional parameters from the nearest similar catchment. The adopted regionalisation procedure is described in detail in Caballero and Rahman (2013). In addition, the above procedure is summarized in Figure 2-2.

MODEL TESTING METHOD

The applicability of the regionalised stochastic inputs and parameters for the state of New South Wales was evaluated by comparing the derived flood frequency curve (DFFC) from the adopted JPA/MCST with the flood quantiles from the at-site FFA and the DEA. The at-site FFA estimates were obtained using FLIKE software (Kuczera, 1999). Here, the at-site observed annual maximum (AM) flood series data were fitted with the Bayesian-Log-Pearson Type 3 distribution. The 90% confidence limits are calculated for the the at-site FFA. The adopted DEA uses the design rainfall intensity data obtained from Australian BOM website following the approach of ARR 1987 (Institution of Engineers Australia, 1987). This adopts the median values of IL (22.30 mm), CL (1.45 mm/h) and *k* (17.54 h) from the calibration of the adopted runoff routing model. The design TP data from Volume 2 of ARR 1987 are also adopted in the application of the DEA.

RESULTS

A total of 19,718 complete storm events were selected from 86 pluviograph stations in the state of New South Wales. This represents an average of 229 events per station and five events per year. The D_{CS} data of the selected complete storm events had an average of 26 h (range: 14 to 55 h) with regional average values of mean and standard deviation of 26 h and 22 h, respectively. The Intensity-Frequency Distribution (IFD) data corresponding to the selected complete storm events had an average of 70 days (range: 42 to 148 days) with regional average values of the mean and standard deviation of 70 days and 83 days, respectively. The IFD tables were developed for each of the selected pluviograph stations following an approach similar to Rahman et al. (2002a). The non-dimensional TP data were also derived for each of the 86 pluviograph stations. For the estimation of IL, CL and *k*, a total of 346 events were selected from the selected six study catchments.

It is found that the D_{CS}, IED, IL and *k* data can be approximated by a gamma distribution and the CL data by an exponential distribution. The D_{CS}, IED and IFD data at any arbitrary location in New South Wales were estimated using the inverse distance weighted averaging method as mentioned earlier. The D_{CS}, IED and IFD data for the selected test catchment (Orara River) were obtained using the D_{CS}, IED and IFD data from the individual D_{CS} and IED data, and IFD curves of pluviograph stations 59026 (located within the catchment), 59040 and 59067, with distances of 3.46 km, 13.67 km and 25.68 km from the centre of the catchment, respectively. Here, the regional TP data were obtained by pooling dimensionless

TP data of 20 pluviograph stations located within 200 km from the test catchment. The IL, CL and *k* data obtained from the regional distributions.

In the simulation, 30,000 sets of input variables/parameters were generated, which were used to simulate 30,000 streamflow hydrographs for the Orara River catchment. The peaks of these simulated hydrographs were used to obtain the DFFC as shown in Figure 2-3. The method of calculating ARIs of the simulated flood peaks can be found in Caballero and Rahman (2013). It can be seen in Figure 2-3 that the obtained DFFC is a good representation of the observed floods up to 10-years ARI. In the upper part, the DFFC shows an under-estimation. However, it should be noted that the observed flood peaks greater than 10 years ARI for the Orara River catchment have high Rating Ratio (RR), which implies a high degree of uncertainty with the observed flood data at higher ARIs. According to Haddad et al. (2010), a large RR indicates that the observed flood data points at higher ARIs are associated with a higher degree of uncertainty and should be used with caution in model validation. Figure 2-3 also shows that the DEA underestimates the smaller ARI floods significantly.

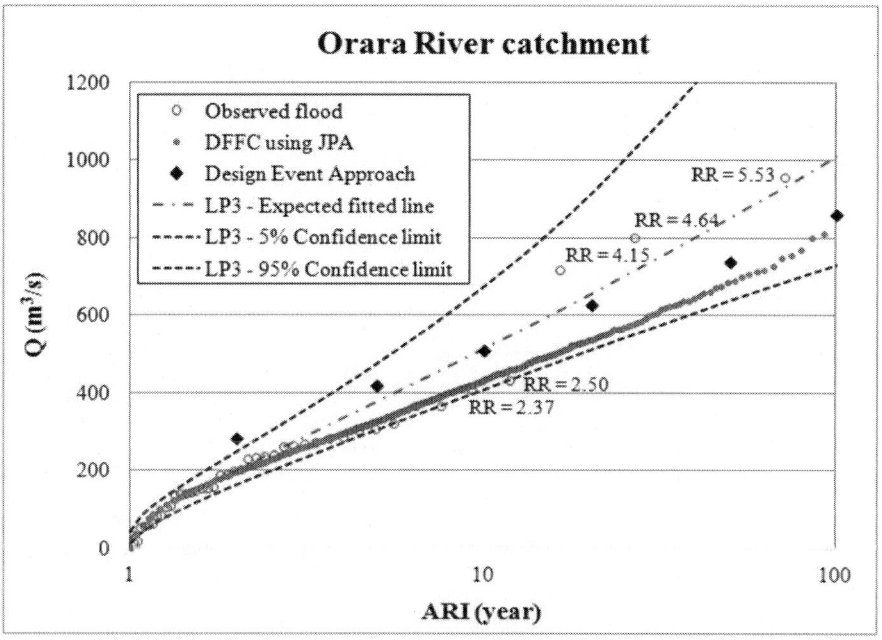

Figure 2.3. Derived flood frequency curve (DFFC) for the Orara River catchment in New South Wales, Australia.

Conclusion

This chapter has found that the Design Event Approach of design flood estimation, currently recommended in Australia, has some serious limitations and hence a more holistic approach should be adopted in practice. It has been found that the stochastic model input variables and parameter values can be regionalised to facilitate the application of JPA/MCST in practice. For the case of New South Wales in Australia, it has been found that the gamma distribution can be adopted to specify the distributions of the rainfall complete storm durations, inter-event durations, initial loss and storage delay parameter of the adopted runoff routing model, while the continuing loss by an exponential distribution. Using an inverse distance weighted averaging method, the regional rainfall intensities at any arbitrary location in New South Wales can be obtained based on the individual IFD tables of the nearby stations. The developed regional TP database can be used to simulate stochastic TP at any arbitrary location in New South Wales.

The regionalised stochastic input variables and parameter data derived for New South Wales were applied to the Orara River catchment for estimating design floods using JPA/MCST. It has been found that the JPA/MCST can provide quite accurate design flood estimates. The regionalised input data developed here with the JPA/MCST although primarily applicable to New South Wales in Australia, it can easily be adapted to other Australian states and countries.

ACKNOWLEDGMENTS

The authors would like to thank the Australian Bureau of Meteorology for providing the pluviograph data and the New South Wales Office of Water (Australia) for providing streamflow data. We also thank the reviewers for their constructive comments which have helped to improve the chapter.

REFERENCES

Aronica G., Candela A. (2007). Derivation of flood frequency curves in poorly gauged Mediterranean catchments using a simple stochastic hydrological rainfall-runoff model. *Journal of Hydrology*, 347(1), 132-142.

Boyd M. J., Bates B. C., Pilgrim D. H., Cordery I. (1987). WBNM: A general run-off routing model. *Fourth national local government engineering conference*. Institution of Engineers Australia, National Conference Publication No. 87/9, 137-141.

Caballero W. L., Taylor M., Rahman A., Shrestha S. (2011). Regionalization of intensity-frequency-duration data: A case study for New South Wales. *The 19th International Congress on Modelling and Simulation*, 12-16 December 2011, Perth, 3775-3781.

Caballero W. L., Rahman A. (2013). Development of regionalized joint probability approach to flood estimation: A case study for Eastern New South Wales, Australia. *Hydrological Processes*. DOI: 10.1002/hyp.9919

Carbone D., Hanson J. (2013). Floods: 10 of the deadliest in Australian history. *Australian Geographic*, viewed on 18th July, 2013. http://www.australiangeographic.com.au/journal/the-worst-floods-in-australian-history.htm.

Carroll D. G., Rahman A. (2004).Investigation of sub-tropical rainfall characteristics for use in the Joint Probability Approach to design flood estimation. *In Proceedings of Second Asia Pacific Association of Hydrology and Water Resources (APHW) Conference*, Singapore, July 2004.

Carroll D. G. (2007). URBS – A catchment runoff-routing and flood forecasting model. User manual.

Charalambous J., Rahman A., Carroll D. (2013). Application of Monte Carlo simulation technique to design flood estimation: A case study for North Johnstone River in Queensland, Australia. *Water Resources Management*, 27, 4099-4111.

Centre for Research on the Epidemiology and Disasters. (2012). Disaster Data. A Balanced Perspective. Issue 27, February 2012, *Centre for Research on the Epidemiology and Disasters (CRED), Universite Catholique de Louvain*, viewed on 18th July 2013, http://www.cred.be/sites/default/ files/CredCrunch27.pdf

FitzGerald G., Du W., Jamal A., Clark M., Hou X. Y. (2010). Flood fatalities in contemporary Australia (1997–2008). *Emergency Medicine Australasia*, 22(2), 180-186.

Gioia A., Iacobellis V., Manfreda S., Fiorentino M. (2008). Runoff thresholds in derived flood frequency distributions. *Hydrology and Earth System Sciences*, 12(6), 1295-1307.

Haberlandt U., EbnervonEschenbach A. D., Buchwald I. (2008). A space-time hybrid hourly rainfall model for derived flood frequency analysis. *Hydrology and Earth System Sciences*, 12(6), 1353-1367.

Haddad K., Rahman A. (2011). Regionalisation of storm duration for determining derived flood frequency curve: A case study for Victoria in Australia. *Asian Journal of Water, Environment and Pollution*, 8 (3), 37-46.

Haddad K., Rahman A. (2012). Regional flood frequency analysis in eastern Australia: Bayesian GLS regression-based methods within fixed region and ROI framework: Quantile Regression vs Parameter Regression Technique. *Journal of Hydrology*, 430-431, 142-161.

Haddad K., Rahman A., Weinmann P. E., Kuczera G., Ball J. E. (2010). Streamflow data preparation for regional flood frequency analysis: Lessons from south-east Australia. *Australian Journal of Water Resources*, 14 (1), 17-32.

Hill P. I., Mein R. G. (1996). Incompatibilities between storm temporal patterns and losses for design flood estimation, 2ndedn. *Hydrology and Water Resources Symposium*, Hobart, 445-451.

Iacobellis V., Gioia A., Manfreda S., Fiorentino, M. (2011). Flood quantiles estimation based on theoretically derived distributions: regional analysis in Southern Italy. *Natural Hazards and Earth System Science*, 11(3), 673-695.

Institution of Engineers Australia. (1987). Australian Rainfall and Runoff - A guide to flood estimation. Volumes 1 and 2. *Institution of Engineers*, Australia.

Jonkman S. N., Vrijling J. K. (2013). Loss of life due to floods. *Journal of Flood Risk Management*, 1(1), 43-56. DOI: 10.1111/j.1753-318X.2008.00006.x

Kjeldsen T. R., Svensson C., Jones D. A. (2010). A joint probability to flood frequency estimation using Monte Carlo simulation. *In proceedings of the BHS Third International Symposium: Role of Hydrology in Managing Consequences of a Changing Global Environment*, Newcastle University, Newcastle Upon Tyne, United Kingdom, July 2010.

Kuczera G. (1999). Comprehensive at-site flood frequency analysis using Monte Carlo Bayesian inference. *Water Resources Research*, 35 (5), 1551-1558.

Kuczera G., Lambert M., Heneker T., Jennings S., Frost A., Coombes P. (2006). Joint probability and design stormsat the crossroads. *Australian Journal of Water Resources*, 10(1), 63.

Laurenson E. M., Mein R. G., Nathan R. J. (2005). RORB Version 5 Rainfall Routing Program – User Manual. Monash University, Department of Civil Engineering in conjunction with Sinclair Knight Merz P/L.

Loukas A., Quick M. C., Russell S. O. (1996). A physically based stochastic-deterministic procedure for the estimation of flood frequency. *Water Resources Management*, 10(6), 415-437.

Loveridge M., Rahman A. (2012). Probabilistic losses for design flood estimation: A case study in New South Wales, *Proceedings of the 34thHydrology and Water Resources Symposium*, Engineers Australia, Sydney, Australia 19-22 November 2012.

Merz R., Bloschl, G. (2004). Regionalisation of catchment model parameters, *Journal of Hydrology*, 287, 95-123.

Mirfenderesk H., Carroll D., Chong E., Rahman M., Kabir M., Van Doorn R., Vis S. (2013).Comparison between design event and joint probability hydrological modelling. *Flood Management Association National Conference*, Tweed Heads, New South Wales, Australia.

Muzik I. (2002). A first-order analysis of the climate change effect on flood frequencies in a subalpine watershed by means of a hydrological rainfall–runoff model. *Journal of Hydrology*, 267(1), 65-73.

Nathan R., Weinmann P. E., Hill P. (2003). Use of Monte Carlo simulation to estimate the expected probability of large to extreme floods, In Proceedings of *28th Hydrology and Water Resources Symposium,* 10-14 November. Wollongong, 1, 105 – 112.

Nathan R. J., Weinmann P. E. (2013). Australian rainfall and runoff discussion paper: Monte Carlo simulation techniques, *Australian Rainfall and Runoff discussion paper*, May 2013, Engineers Australia.

National Committee on Water Engineering (2013). Australian Rainfall and Runoff revisions projects, *National Committee on Water Engineering*, Engineers Australia, http://www.ncwe.org.au/arr/index.html.

Price Waterhouse Cooper (2011). Economic impact of Queensland's natural disasters. *Price Waterhouse Cooper*, http://www.pwc.com.au/about-us/flood-support/assets/Economic-Impact-Qld-Natural-Disasters.pdf

Rahman A., Weinmann P., Hoang T., Laurenson E. (2002a). Monte Carlo simulation of flood frequency curves from rainfall. *Journal of Hydrology*, 256(3), 196-210.

Rahman A., Weinmann E., Mein R. G. (2002b). The use of probability-distributed initial losses in design flood estimation. *Australian Journal of Water Resources*, 6(1), 17.

Svensson C., Kjeldsen T. R., Jones D. A. (2013). Flood frequency estimation using a joint probability approach within a Monte Carlo framework. *Hydrological Sciences Journal*, 58(1), 8-27.

Sih K., Hill P., Nathan R. (2012). Joint probability model for the Nerang River catchment stage II. Report prepared for the Gold Coast City Council by SKM, Melbourne.

Vaze J., Perraud J. M., Teng J., Chiew F., Wang B. (2011). Estimating regional model parameters using spatial land cover information - implications for predictions in ungauged basins. *19th International Congress on Modelling and Simulation*, Perth, Australia, December 2011, 3433-3440.

Viglione A., Blöschl G. (2009). On the role of storm duration in the mapping of rainfall to flood return periods. *Hydrology and Earth System Sciences*, 13(2), 205-216.

Weinmann P. E., Rahman A., Hoang T., Laurenson E. M., Nathan R. J. (1998). A new modelling framework for design flood estimation. *Proceedings of International Symposium on Stormwater Management*, 27-30 Sept, Adelaide, 1, 393-398.

Weinmann P. E., Rahman A., Hoang T., Laurenson E., Nathan R. J. (2002). Monte Carlo simulation of flood frequency curves from rainfall-the way ahead. *Australian Journal of Water Resources*, 6(1), 71-80.

In: Flooding
Editors: M. R. Motsholapheko and D. L. Kgathi

ISBN: 978-1-63117-607-4

Chapter 3

CAUSES AND IMPACTS OF FLOODS IN THE NORTHEAST OF THE PROVINCE OF BUENOS AIRES, ARGENTINA

***E. E. Kruse*[1], *J. L. Pousa*[1,*] *and D. O. Guaraglia*[2,†]**

[1]Consejo Nacional de Investigaciones Científicas y Técnicas (CONICET), Argentina. Facultad de Ciencias Naturales y Museo, Universidad Nacional de La Plata, La Plata, Argentina

[2]Consejo Nacional de Investigaciones Científicas y Técnicas (CONICET), Argentina. Facultad de Ingeniería, Universidad Nacional de La Plata, La Plata, Argentina

ABSTRACT

The northeast region of the Province of Buenos Aires, Argentina comprises a coastal plain along the Rio de la Plata estuary and an adjacent continental plain.

This region exhibits a high degree of urbanization associated with continuously growing economic activities.

Regional plain features, wet climate, occurrence of extra-tropical storm surges, and land changes carried out without suitable planning have often produced periodic and disastrous floods in highly urbanized areas particularly those located in topographically low-lying sections. Heavy rainfall events of progressively increasing frequency, the diminishing of infiltration due to extensive urbanization and the subsequent increment in surface runoff have aggravated the problem. This chapter aims at analyzing the causes and environmental impacts of flooding in Province of Buenos Aires. The tragic flooding of the 2nd April 2013 in La Plata has been selected as a paradigmatic case study. Hydro-meteorological monitoring for three locally possible flooding scenarios is considered. A series of structural and non-structural measures that would facilitate the mitigation of

---

* Corresponding author: J. L. Pousa. Consejo Nacional de Investigaciones Científicas y Técnicas (CONICET), Argentina. Facultad de Ciencias Naturales y Museo, Universidad Nacional de La Plata, Calle 60 y 122, (1900) La Plata, Argentina. E-mail: jlp@fcnym.unlp.edu.ar.

† D. O. Guaraglia: Consejo Nacional de Investigaciones Científicas y Técnicas (CONICET), Argentina. Facultad de Ingeniería, Universidad Nacional de La Plata, Calle 1 y 47, (1900) La Plata, Argentina.

flood damage were also analysed. Such measures are recommended for implementation in the continental and the coastal plains.

Keywords: flooding, coastal plain, continental plain, extreme rainfall, storm surges, Río de la Plata

INTRODUCTION

With a population that is now over 13 million, above one third of the total population of Argentina, the northeast region of the Province of Buenos Aires exhibits a high degree of urbanization associated with continuously growing industrial, commercial and agricultural activities. Approximately 50% of the Gross Domestic Product (GDP) is generated in this region, considered as the most dynamic in Argentina. Geographically, the area comprises coastal zones on the Rio de la Plata estuary and continental zones associated with drainage basins that discharge into the estuary (Figure 3-1).

The continuous and rather chaotic growth of the region during the last century, together with its own socio-economic features, have resulted in the existence of many problems related to water resources, one of them referred to as periodical flooding. The lack of suitable land-use planning that takes into account the characteristics of the physical environment has led to the occupation of low-lying, flood-prone areas, thus aggravating the consequences of flooding. Floods have become a crucial question with respect to the possibilities of future socio-economic development of the region.

This chapter aims at analysing the causes and impacts of flooding in the north-east region of the Province of Buenos Aires, with special emphasis on recent events that occurred in the city of La Plata and its surroundings.

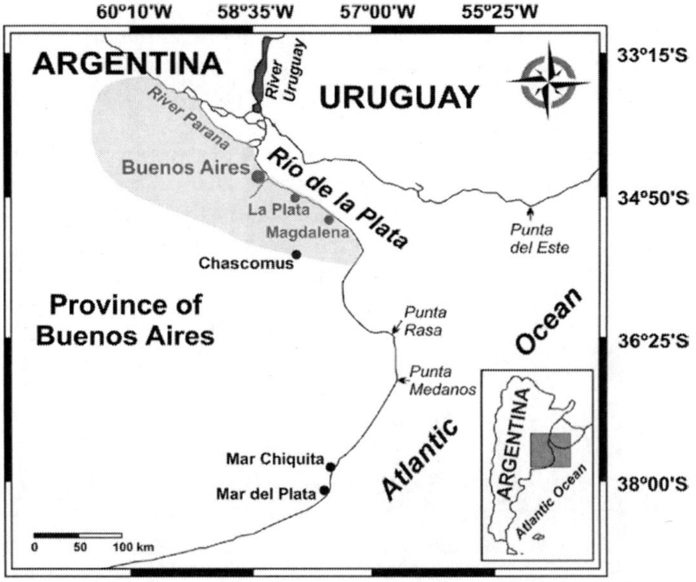

Figure 1. Location map. The light shaded area corresponds to the northeast region of the Province of Buenos Aires.

These events resulted in great repercussions at an international level because of their social and environmental impacts.

REGIONAL SETTING

The north-east region of the Buenos Aires Province, particularly the city of La Plata and its surroundings, is characterised by a plain landscape with mild undulations where surface runoff from watersheds of generally little extension discharges into the Rio de la Plata. Two contrasting environments can be defined on a regional scale: the continental plain and the coastal plain (Figure 3-2). The limit between both plains is located at about 5 m over mean sea level (MSL). This limit is given by the marine ingression in the Holocene.

Fluvial basins with well-defined courses develop in the continental plain. They have a moderate to good integration into the drainage network. As watersheds acquire their own characteristics, the main course becomes more important until it reaches the middle and lower sections, which have low banks which separate the main course from its floodplain.

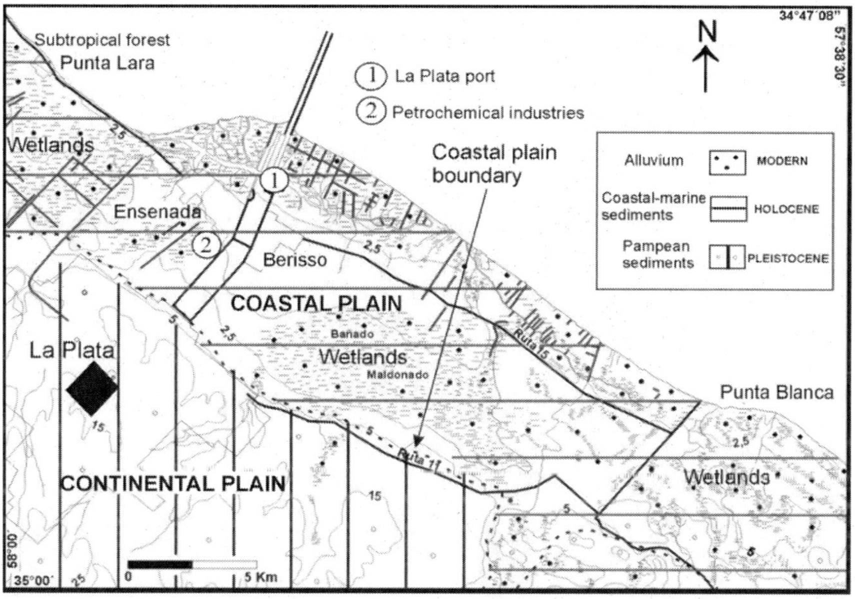

Figure 2. Geologic and topographic outline of the area around La Plata, showing the coastal plain features. Note the width of the coastal plain that has been occasionally affected by extreme surges.

A generalised characteristic of the drainage basin is that rivers and creeks are gaining streams (or effluent streams) because they gain water from local aquifers. The lithologic composition of this environment corresponds to clayey silts, known as Pampian loess.

The coastal plain develops from the height of 5 m to the present coastline. This plain is an extended fringe, which is 4-8 km in width and it is parallel to the Rio de la Plata, characterized by a very low slope (from 0.06 to 0.1%), with a hydrographic network, represented by a few main watercourses and tributaries, in an environment that exhibits a chaotic drainage design and extended depressed swamps. The monotony of this relief is locally altered by the presence of small hills, not more than 2-2.5 m high, developed parallel to the shoreline.

In regional terms, and according to the Köppen climate classification, the north-east of the Province of Buenos Aires is humid temperate. Mean annual precipitation is about 1,000 mm, but there are dry and humid cycles that strongly alter the average values (Kruse and Laurencena, 2005). A humid climatic cycle, characterized by an increase in rainfall that has exceeded the historic records, has become evident since 1970.

As a result of this rainfall increase, there have been frequent changes in the hydrologic regime of the region that gave rise to higher water table levels and numerous flooded areas in a variety of degrees. Paleoclimatic changes have left many important signs in the geological features of the region, the most detectable of them being, among others, palaeosols (humid periods) and cyclic dune formation (dry periods).

Therefore, the nature of the region allows significant climatic oscillations to be foreseen. In this regard, there are historic references that show the existence of alternating dry and humid periods. Two examples of these alternating periods are the locally so-called "gran seca de Darwin" (the great Darwin's dry) between 1827 and 1832, and the arrival at Chascomus (Figure 2-1) in 1857 of a steamer that had departed from Buenos Aires during a humid period (Maiola et al., 2001).

Land misuse, groundwater overexploitation, and effluent and waste disposal have led to significant modifications of the hydrologic cycle. On the whole, these practices have reduced the natural infiltration of rainfall, with the subsequent decrease of evapo-transpiration and the increase of surface runoff. Heavy rains in urbanized zones can give rise to considerable effects because infiltration diminishes and surface runoff increases. The problem is aggravated when people settle in the floodplains of rivers and creeks, as has often been the case.

The main problem in the continental plain comes from periodic flooding due to extreme precipitation with a rapid surface runoff towards urbanized sectors. In the coastal plain, instead, the main natural events that threaten inhabitants are the overflowing of the Rio de la Plata shores due to southeasters, and the flooding generated by rainfall and overflowing of streams and canals from the continental sector.

Flooding Events

The occurrence of storm surges (southeasters) and heavy rainfall can give rise to extreme flooding. Storm surges have their greatest impact on the coastal plain, particularly if they bracket several tidal cycles, whereas heavy rainfall is most problematic on the continental plain.

Such meteorological events associated with a regional plain feature, a wet climate and land changes of anthropic origin carried out without a suitable planning, have produced periodic floods with severe socio-economic impacts which very often have resulted in disastrous consequences.

Tides and Storm Surges

Changes in the water level in the Rio de la Plata estuary are due to the co-oscillating tide forced by the global tide oscillation along the edge of the continental shelf (O'Connor, 1991). The resultant tide in the estuary is mixed, predominantly semi-diurnal. The principal semi-diurnal lunar tide (M2) has amplitude of 0.27 m at Buenos Aires. However, successive low waters and high waters exhibit large diurnal inequalities due basically to the principal diurnal declinational lunar tide (O1) whose amplitude is 0.15 m (D'Onofrio et al., 1999). Average tidal ranges are 0.61 m at Buenos Aires, 0.54 m at the La Plata harbor and 0.80 m at Punta Rasa (Figure 3-1) (Servicio de Hidrografia Naval [SHN], 2013). However, because of its large extension, shallow depth and funnel-like shape, the Rio de la Plata estuary is highly affected by extratropical storm surges that push its waters upstream, causing severe flooding on the Argentine shore. Extratropical storm surges can be generated either in the southern portion of the Argentine continental shelf or in the maritime front of the Province of Buenos Aires (Lanfredi et al., 1998; Schnack and Pousa, 2002; Pousa et al., 2007; Schnack et al., 2010).

Apart from the distinction between tropical and extratropical, storm surges can be classified as positive, when they raise the water level over the astronomical tide, or negative, when they reduce the water level below the predicted tide. Positive surges, together with high tides and waves, can cause severe coastal flooding, whereas negative surges can be potentially dangerous for shipping.

Although positive surges in the Rio de la Plata estuary are the main subject of this chapter, negative surges will be briefly mentioned in connection with the simultaneous occurrence of both types of surges in a relatively short distance along the coast of the Province of Buenos Aires.

Many extratropical positive storm surges have been recorded along the Argentine coast simultaneously with the northward traveling tidal wave. They are basically produced by the combined action of an anticyclone located to the west of Argentina (semi-permanent Pacific anticyclone) and a cyclone located over the Atlantic to the southeast of Argentina, the latter moving towards the east or northeast. Because of this situation, strong winds from the south or southwest (clockwise circulation around a low in the southern hemisphere) and high water levels affect the whole coast of Argentina, as well as the Rio de la Plata shores, Uruguay, and even southern Brazil.

However, the most conspicuous and worst floods along the Rio de la Plata shores are due to southeasters (locally known as "*sudestadas*").

These are very strong winds from the SE-SSE, often caused by an anticyclone located over southern Argentina and the adjacent ocean (anti-clockwise circulation around a high in the southern hemisphere). These winds are accompanied by a sky completely covered by nimbostratus and persistent rainfall (SHN, 1999). Southeasters last from several hours to two or three days, thus bracketing several tidal cycles. But in order that water might attain an extraordinarily high level, it is necessary, among other factors that a depression forms to the north of Buenos Aires, and over Uruguay and southern Brazil. Figure 3-3 illustrates the weather chart corresponding to the flood of 15 April 1940, when the predicted tidal level in Buenos Aires (1.20 m) was overcome by 3.24 m, giving the highest water level in Buenos Aires (4.44 m over datum) since the beginning of records in 1905. Twenty five people were killed. Floods of similar characteristics were experienced in 1958, 1959 and 1989, and more recently in 1993 and 2001.

Every time this phenomenon occurred, it resulted in severe property loss and other damages. Coastal plain flooding due to storm surges can be particularly destructive in the Rio de la Plata shores because their topographic gradients are extremely low. It is not difficult to guess the impact that these meteorological events can have on the low-lying estuary shores. Figure 3-4 shows an example of the 21 March 2001 storm surge on the Rio de la Plata coastal plain near La Plata. At 17:00 (local time) the observed water level was 1.83 m over the predicted tide. Flooding is exacerbated by the Coriolis force piling-up the waters as storm surges travel along the Argentine coast, particularly within the Rio de la Plata.

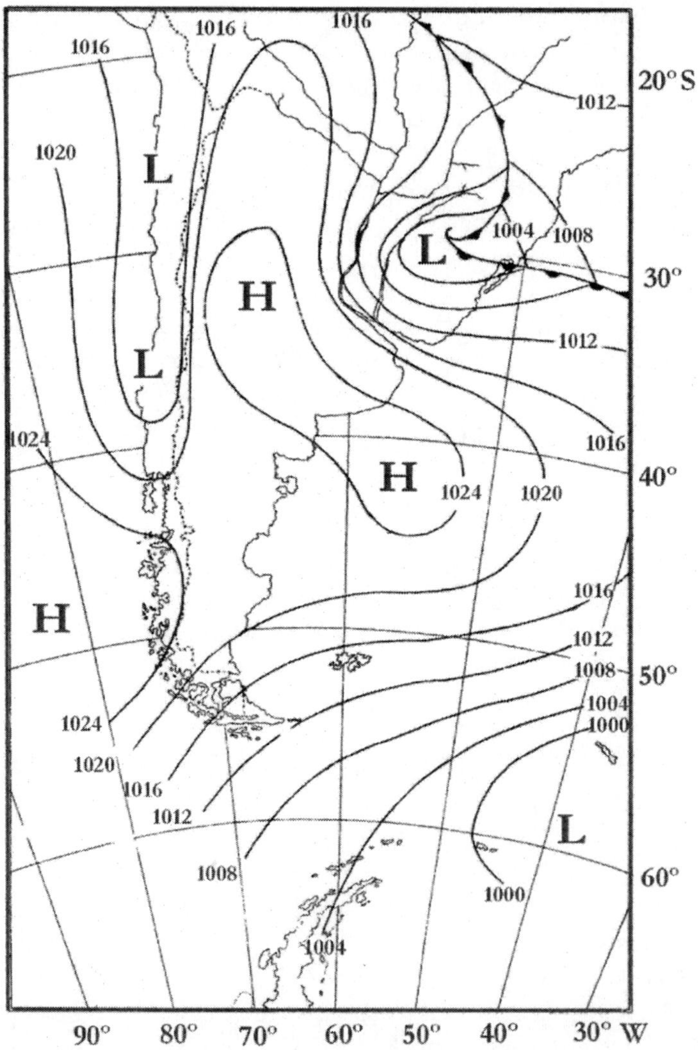

Modified from Balay, 1961.

Figure 3. Synoptic weather chart of the severe storm surge event of 15th April 1940.

According to Pugh (1987), both negative and positive surges may be generated by the same storm at different stages of its progression.

Nevertheless, a remarkable point about the Province of Buenos Aires shores is that the same storm event can simultaneously give rise to a positive surge on the maritime front and a negative surge in the Rio de la Plata estuary (Pousa et al., 2013). A relatively recent example of this is the 21–23 July 2009 storm surge.

On that occasion, a severe positive storm surge affected almost the entire 5,000 km coast of Argentina. This storm event produced a very different situation at Buenos Aires and other nearby coastal sites along the west margin of the Rio de la Plata. Blowing this time from the southwest (recall clockwise circulation around a low in the southern hemisphere), the winds swept the estuary waters away from the coast. Thus, the same positive storm surge that battered the entire Atlantic coast piling up the waters along it became, at the same time, a negative storm surge in the Rio de la Plata.

Rainfall

Floods in the continental zone are related to hydrographic basins of little areal development and low drainage density.

Heavy rain events of progressively increasing frequency, along with an increment in surface runoff, have produced significant flooding in urbanized areas located in topographically low-lying sectors or settled in the floodplains of the major local streams. From a hydrological point of view, these factors connected with the global climate change, namely the relative sea-level rise and changes in the hydrological cycle due to changes in temperature and rainfall, put forward more critical scenarios concerning the intensity and frequency of floods.

Precipitation has exhibited an increasing trend in recent decades, a factor that should be taken into account in any undertaking involving environmental consequences.

Data from regional meteorological stations show that precipitation has increased from 900 mm/year at the beginning of the series in 1910 to 1100 mm/year at the end of the series in 2010. In the case of the warm season (October – March), rainfall figures range from 700 mm at the beginning to 900 mm at the end of the series. Trends are similar for the cold season (April – September), but with smaller values than in the warm season. The highest figures for precipitation are obtained during the period February-April and October-December., Storms occur mainly during the periods March-May and August-October, and they are predominantly associated with the interaction between cold and warm fronts.

Rainfall dynamics is governed by the encounter between southwest cold and dry air masses from the huge South Pacific anticyclone, with warm and humid air masses from the South Atlantic anticyclone. The region is affected by storms 50 days a year on average. Convective precipitation from vertically developed clouds is heavy and of short duration.

Flooding is a direct result of particularly high pluviometric intensity. The development of convective clouds is due to unstable atmospheric conditions that generate suitable conditions for the upward movement of air masses (convective development) from intermediate atmospheric levels. This type of precipitation develops not only during summer, but also during autumn and spring. In fact, heavy rainfall of short duration has increased its frequency in recent times. Table 3-1 shows rainfall events in La Plata which have produced severe flooding. All of them correspond to a 24-hour precipitation.

Photo courtesy of El Dia newspaper, La Plata.

Figure 4. The storm of 21st March 2001 striking the seawall and coastal road at Punta Lara, the coastal zone of La Plata (Figure 1).

Case Study: The Flooding of La Plata on 2nd April 2013

An extraordinary rainfall of 392 millimeters in 24 hours resulted in flash flooding on the 2nd April 2013, in La Plata, caused the collapse of most of the city and resulted in property losses and evacuation of thousands of people. More than 100 people died (Figure 3-5).

The rainfall amount exceeded the mean monthly values in historic records. Meteorologically, this event of very heavy rainfall was characterized by the presence of a depression in middle levels of the troposphere. These systems are more frequent during autumn, and affect the central zone of Argentina with an average of three rainfall events in autumn. However, their intensity is variable and it is associated with precipitation in the region. The same applies to other meteorological situations; it is still difficult to accurately forecast the intensity location and time of occurrence of these extreme rainfall events. It is a matter that requires further investigation at the regional level.

Table 3.1. Rainfall figures in La Plata, 1911 to 2013

Date	Precipitation (mm)
24 April 1911	186.0
10 March 1930	225.0
6 March 1962	131.0
14 May 1980	155.1
5 January 1998	119.4
20 November 1999	140.0
27 January 2002	121.2
28 February 2008	240.0
2 April 2013	392.2

Source: Faculty of Engineering (National University of La Plata).

Figure 5. Two photographs of the 2nd April 2013 flood in La Plata.

The drainage network of La Plata has been largely modified by urbanization. There is practically not any fluvial course that does not show some degree of anthropization. Furthermore, some courses have apparently disappeared. As a consequence of this extreme event, the runoff regimes of the creeks with basins located within the city were largely exceeded. Water quickly occupied its old courses, now entirely urbanized, and extended towards floodplains, thus flooding many sectors of the city. In addition, the surroundings of La Plata, not so long ago almost uninhabited, have become a fruit and vegetable belt, quite urbanized and with many greenhouses, which reduced infiltration and increased runoff. Because of this, huge quantities of rainwater from the surroundings of La Plata entered the city, thus rapidly aggravating an already troublesome situation.

The urban residential stormwater drainage system was of little efficiency in the face of such an extreme event. The flood took a toll of more than 200,000 inhabitants gravely affected (with more than one hundred people dead) and severe property losses.

These consequences stress the need for a proper management of water resources supplemented with non-structural measures for the prevention of extraordinary events. The intensity of rainfall was unusual, and the existence of highly urbanized zones within the floodplains of local streams largely aggravated the impacts of this exceptional storm event. However, lack of an integrated management of flood risk is a significant factor in analyzing the consequences of the meteorological event. It was a determinant factor in the absence of the implementation of preventive and corrective actions, and also critical during the emergency.

HYDROMETEOROLOGICAL MONITORING

General Features

From the point of view of monitoring, three possible scenarios can be recognized with the aim of warning about an impending flood. One of the scenarios refers to slow flooding in scarcely populated areas. With respect to the study area, this happens when prolonged rainfall combines with a rise in the Rio de la Plata level, thus flooding nearby rural zones. This flooding can have significant economic consequences, but human losses use to be low.

Another scenario is when the meteorological event has similar characteristics as in the first case but reaches urban areas densely populated. Such a situation, with a gradual advance of flooding, allows local authorities to display measures for rescuing and moving people to more safe places, thus mitigating the impact of losses, particularly human losses.

The worst flooding scenario occurs when it takes place in densely populated areas and a heavy rain falls within few hours. In this case the local authorities have no time in which to prepare and organize a response, nor can they go to the rescue of inhabitants, so human losses can be high.

In the Province of Buenos Aires, there are monitoring systems whose data are used for weather forecasting. There are also several water level gauges along the Rio de la Plata which are basically used for supporting shipping and nautical sports. This information is also used to estimate the probability of flooding during southeasters.

As the goal is to collect meteorological parameters at a regional level, the location of instruments and the spatial and temporal system scales, are not generally adequate for dealing with the last kind of flooding situations.

In the first two scenarios, in which the development of flooding is slow (several days), it is possible to use the already existing monitoring system to limit property losses and execute an evacuation of people. Few level measurements in the Rio de la Plata, as well as some rainfall measurements combined with visual reports from local police forces or from inhabitants themselves, will be enough for this purpose.

As stated above, the worst situation is that of the rapid flooding of the densely populated areas. In these cases traditional monitoring systems are practically useless. This situation produces severe human and property losses because people cannot be moved rapidly to other less dangerous places. It should be dealt with by local authorities with more funds to lessen the flooding impacts through the adoption of new monitoring systems.

Probability forecasting of flooding like that of 2nd April 2013 in La Plata, requires particular monitoring tools and a computing model running in real time and fed by permanently updated data. This system could allow the authorities to respond immediately to flooding impacts.

Selecting the elements for this kind of monitoring requires some details of the problem to be understood. Although cities, towns and villages within the study area are located in a plain (the northeast sector of the Pampian plain), the relief within them gradually changes within distances of a few hundred metres. Therefore, it is worth noting that even when rainfall can be uniform over the whole site, the topographic features produce preferential flow, which is a rapid movement of water that can reach high velocities. These characteristics should be taken into account when dealing with the selection of places to install the monitoring instruments, and also with the feature of the instruments itself because they should have a fast response to flooding events.

Because catastrophic flooding occurs in a reduced area (a town) and in few hours, the monitoring system should have spatial and temporal resolutions much greater than those offered by traditional monitoring systems.

A Solution for the Monitoring Problem

Monitoring is the first stage of an early warning system for flooding, but it should be used with a hydrological model that simulates the local problem. This model should be fed with data from the monitoring network.

After a reasonable time has elapsed, in which several rainfall events have occurred, the model might be progressively adjusted in order for it to predict flooding. It is a "to and fro process" in which through the use of the model the need for new measurements, not previously taken into account, could be detected. Some guidelines will be developed for a generic case of monitoring in a floodable city with the characteristics described in this chapter.

As a first step, the system should take full advantage of the information collected by the traditional monitoring system stations already existing within the zone.

Data produced by these instruments should be reported to a management center through some safe and rapid communication means. The predictive model will be run at this center. The collection of data on possible flooding events should be done with a much higher frequency than the collection of data for purely meteorological purposes. Perhaps it would be adequate to begin with a periodicity of one minute and as the model is used the period can be adjusted.

In order to determine which instruments will compose the monitoring system and their location, it is necessary to have knowledge of the watershed which should be supplied by hydrogeologists. Pluviographs should be installed at the feeding points of the watershed to have a continuous rainfall assessment. As rainfall usually has an heterogeneous distribution, and the zones toward water runs depend, partially, on which are the zones of rainfall, it is necessary an important network of suitably distributed pluviographs. These pluviographs may be of the tipping bucket type, but because in predicting flooding it is very important to determine rainfall intensity, they must allow each lever tip to be instantaneously recorded and informed.

Water level should be particularly measured in those low-lying points suggested by researchers according to previous experiences. Not only should water level in the Rio de la Plata be measured, but also in the regional creeks that collect water from the whole basin.

All these level loggers could be pressure sensors. We have successfully used submersible pressure sensors with the reference port referred to a given constant pressure. When using these sealed sensors, the varying atmospheric pressure is superimposed to the water column pressure. Therefore, to evaluate the exact level the atmospheric pressure must be subtracted.

This is simple to do in an area that includes the city and its surroundings, because only one pressure measurement would be enough to compensate all the gauge sensors.

Submersible pressure sensors which compensate for local atmospheric pressure changes can also be used.

These sensors, called *vented gauge sensors,* expose the reference port to the atmospheric pressure by means of a vent tube placed into the same cable used for the electric power supply and output signals. This vent tube would allow atmospheric pressure variations to be compensated, but in practice any obstruction restricting air movement along the vent tube could prevent the atmospheric pressure from reaching the sensing membrane and would thus introduce errors in the measurements. Sometimes humidity condensation is the cause of the vent tube obstruction. In some cases it could be necessary to measure flow in creeks.

Generally, creeks that get across urbanized zones can drag elements potentially dangerous for the integrity of instruments. Perhaps the Acoustic Doppler Current Profiler (ADCP) could be suitable for flow velocity measurements. The word "profiler" in the name of the current meter means that they are able to measure flow velocity at different distances from the transducer, the distances ranging from decimeters to several meters. According to the amount of sensors used, they can be two or three-dimensional profilers.

Measuring water level in channels is much simpler, cheaper and less prone to be damaged than using flowmeters. But in order that level measurements might be a useful tool it is necessary to know the relation *h-Q* for different channel conditions. Therefore, systematic measurements of the relation *h-Q* should be carried out under great discharge conditions along channels. This could also be useful to decide when channels should be weeded or when the section of a canal should be subjected to maintaining operations.

In order to have the monitoring system properly working in critical events, it is of the utmost significance to have a communication system that keeps working even if the zone is completely flooded. Therefore, electronic parts should be hermetically sealed and/or installed in places where the flooding does not affect them. Instruments and communication devices should be fed by batteries and alternative energy sources, as in case of severe flooding the electrical energy is cut off for safety reasons, or simply because water drags part of the energy distribution system.

MITIGATION MEASURES

A series of structural and nonstructural measures would facilitate the mitigation of damage, such as the display of an efficient monitoring system, the construction of structures to buffer the water level rise and the cleaning of riverbeds and watercourses among structural measures, and land-use planning and regulation among nonstructural measures.

To be more specific the mitigation measures can be separately planned for the continental plain and the coastal plain.

Continental plain:

- Promote the creation or the maintenance of green areas as a protection against flooding. A low-cost, flooding-resistant recreational infrastructure can be included in these sectors. These sectors can simultaneously be recreational areas (urban parks) and conservation zones.

- Limit the growth and occupation of potentially floodable zones, particularly on the banks of the creeks. Local authorities should also assess the possibility of relocating certain social groups in less risky sectors.
- Ensure the maintenance of water courses and canals by cleaning and weeding their beds.
- Promote the participation of the local community at all levels by diffusion and awareness campaigns.

Coastal plain:

- Build coastal works, including parks, with flooding-resistant infrastructure in the face of southeasters.
- Limit the occupation of the coastal border and keep restrictions to land use in low-lying, potentially floodable zones (floodplains of creeks, swamps and depressions), thus avoiding new fillings in those sectors.
- Build defense structures adjacent to coastal towns keeping the appropriate height that lessen the impact of flooding due to southeasters.

A measure that would be useful for both the continental plain and the coastal plain is to keep the natural pristine conditions of those highly vulnerable sectors in order that they may be a natural protection against flooding.

CONCLUSION

The northeast region of the Province of Buenos Aires exhibits two geomorphologic environments, the coastal and the continental plain, which are often affected by severe floods as a consequence of different causes.

Flooding in the continental plain is related to short-term extreme rainfall events whose runoff is accelerated because of urbanization and other human activities.

This affects particularly people who have settled near the floodplains of water courses. These events are responsible for significant negative impacts such as heavy economic losses, as well as loss of life in the most populated sectors. This was the case of the catastrophic event occurred in La Plata on 2nd April 2013 with a precipitation higher than 300 mm in only four hours.

In the coastal plain, flooding is mainly connected with positive storm surges that notably increase water level in the Rio de la Plata estuary. Positive storm surges give origin to severe damage for coastal inhabitants and for the infrastructure built in this area.

The development of efficient monitoring systems and the implementation of structural and nonstructural measurements are basic matters to take into account as regards the mitigation of damage from flooding.

REFERENCES

Balay, M. A., 1961. *El Rio de la Plata entre la atmósfera y el mar.* H-621; Servicio de Hidrografía Naval: Buenos Aires.

D'Onofrio, E. E., Fiore, M. M. E., Romero, S. I., 1999. Return periods of extreme water levels estimated for some vulnerable areas of Buenos Aires. *Cont. Shelf Res.* 19, 1681-1693.

Kruse, E., Laurencena, P., 2005. Aguas superficiales. Relación con el régimen subterráneo y fenómenos de anegamiento. In: De Barrio, R. E., Etcheverry, R., Caballé, M. F., Llambías, E. (Eds.). *Relatorio del XVI Congreso Geológico Argentino* (pp. 313-326). Asociación Geológica Argentina. La Plata.

Lanfredi, N. W., Pousa, J. L., D'Onofrio, E. E., 1998. Sea-level rise and related potential hazards on the Argentine coast. *J. Coastal Res.*, 14(1), 47-60.

Maiola, O., Gabellone, N., Hernández, M., 2001. *Inundaciones en la Región Pampeana.* In: Universidad Nacional de La Plata (Ed.); La Plata.

O'Connor, W. P., 1991. A numerical model of tides and storm surges in the Rio de la Plata Estuary. *Cont. Shelf Res.*, 11(12), 1491-1508.

Pousa, J., Tosi, L., Kruse, E., Guaraglia, D., Bonardi, M., Mazzoldi, A., Rizzetto, F., Schnack, E., 2007. Coastal processes and environmental hazards: The Buenos Aires (Argentina) and Venetian (Italy) littorals. *Environ. Geol.*, 51,1307-1316.

Pousa, J. L., D'Onofrio, E. E., Fiore, M. M. E., Kruse, E. E., 2013. Environmental impacts and simultaneity of positive and negative storm surges on the coast of the Province of Buenos Aires, Argentina. *Environ. Earth Sci.*, 68(8), 2325-2335.

Pugh, D. T., 1987. *Tides, Surges and Mean Sea-Level.* John Wiley and Sons: Chichester, UK.

Schnack, E. J., Pousa, J. L., 2002. *Episodic processes.* In: Isla, F. I. (Ed.); Coastal systems, Coastal Dynamics. *Encyclopedia of Life Support Systems (EOLSS).* UNESCO, Oxford, UK. http://www.eolss.net.

Schnack, E. J., Pousa, J. L., Bértola, G., Isla, F. I., 2010. Argentina (Chapter 4.4: 219-225). In: Bird, E. (Ed.); *Encyclopedia of the World's Coastal Landforms*; Springer: Heidelberg, Germany.

Servicio de Hidrografia Naval [SHN], 1999. *Derrotero Argentino, Parte I, Rio de la Plata; H-201;* Buenos Aires, Argentina.

Servicio de Hidrografía Naval [SHN], 2013. *Tablas de Marea*; http://www.hidro.gov.ar/Oceanografia/Tmareas/Form_Tmareas.asp.

In: Flooding
Editors: M. R. Motsholapheko and D. L. Kgathi
ISBN: 978-1-63117-607-4

Chapter 4

MAPPING OF AREAS VULNERABLE TO LEPTOSPIROSIS BASED ON THE ANALYSIS OF FLOODS IN THE SÃO PAULO METROPOLITAN AREA

Andrea Ferraz Young*
King's College London –Department of Geography, Strand, London, UK

ABSTRACT

This chapter focuses on the relationship between climate change and the impacts of climate events, as well the importance to manage the associated risk. According to IPCC (2012), the impacts of climate extremes and the potential for disasters emerge from the interaction of physical conditions (weather/climate), from the exposure and vulnerability of human and natural systems. With global climate change, extreme weather events such as cyclones and floods are expected to occur with increasing frequency and greater intensity, which contaminates freshwater supplies, heightens the risk of water-borne diseases, and creates breeding grounds for diseases such as leptospirosis. Infectious disease outbreaks have been reported following major flood events in developing countries, and these outbreaks vary in magnitude and rates of mortality. The objective of the study was to evaluate leptospirosis at geographic locations based on environmental factors and produce a predictive disease risk map. The area of study is São Paulo Metropolitan Area (SPMA), one of the largest megacities in the world (with 19,956,590 people). In this case, we considered the previous analysis of Coelho and Massad (2012), where for each additional 20 mm of precipitation there was an average increase of 15.6 % in hospital admissions, as the precipitation increased from 20 mm to 120 mm. Within this context, we considered the spatial distribution of the daily number of leptospirosis cases and the probability of flood events in SPMA. We used ArcGIS to integrate the spatial information and non-spatial attribute data, where each spatial feature and its attribute information were linked. The ArcGIS also provided the "math" module to perform mathematical operations in order to analyze the geographical patterns and trends of the region. As a result, a map of vulnerability was produced based on the leptospirosis risk evaluation which can enable decisions to be made on strategies and interventions for improving the conditions of people living in disease-prone areas.

---

* Corresponding Author address: Email: andrea.young@kcl.ac.uk

Keywords: leptospirosis, precipitation, health, and climate change

Introduction

Leptospirosis, which is caused by spirochetes of the genus "Leptospira", is acquired through direct or indirect contact with the urine of infected animals (e.g., rats) (Bharti et al. 2003; McBride et al. 2005). Leptospirosis is a worldwide zoonotic infection with a much greater incidence in tropical regions and was identified as one of the emerging infectious diseases (Levett, 2001).

Leptospirosis has been recognized as an emerging global public health problem because of its increasing incidence in both developing and developed countries (Farr 1995; Levett, 2001; Biosci et al., 2008 Vanasco et al., 2004).

In their article about Samoa, Lau et al. (2012) mention that, "environmental factors are important in determining the risk of human infection, and differ among urban settings." According to Lau et al. (2012) "a wide range of risk factors include high precipitation and flooding; poor sanitation and hygiene; urbanisation and over-crowding; contact with animals (including rodents, livestock, pets, and wildlife); and environmental degradation."

Climatic conditions are strongly connected to water-borne diseases and diseases transmitted by animals (such as rats) during flood events. A number of leptospirosis outbreaks have occurred in the past few years in various places such as Brazil, India and Nicaragua. Some of these resulted from natural calamities such as cyclone and floods (Biosci et al., 2008).

Floods are increasing in frequency and intensity (IPCC, 2007; 2012), which contaminates freshwater supplies, heightens the risk of water-borne diseases, and creates breeding grounds for diseases such as leptospirosis. Infectious disease outbreaks have been reported following major flood events in developing countries, and these outbreaks vary in magnitude and rates of mortality (Lau et al., 2010).

According to IPCC (2007), studies published since the Third Assessment Report (TAR) support previous projections that climate change could alter the incidence and geographical range of infectious diseases.

> "Flooding and heavy rainfall have been associated with numerous outbreaks of leptospirosis around the world. With global climate change, extreme weather events such as cyclones and floods are expected to occur with increasing frequency and greater intensity and may potentially result in an upsurge in the incidence as well as the magnitude of leptospirosis outbreaks" (Lau et al., 2010).

Much of the resurgent international interest in leptospirosis stems from several large clusters of cases which have occurred in Central and South America following flooding as a result of excess rainfall related to El Niño (Levett, 2001). In accordance with Soares et al. (2010), leptospirosis is spread throughout the municipality of São Paulo, and its incidence increases during the rainy season. Flash floods have become common throughout the SPMA during the summer season, which runs from December to March (Nobre et al., 2011).

Few studies analysing leptospirosis from the viewpoint of spatial distribution have been produced (Soares et al., 2010). The use of geo-processing techniques in health studies greatly help to formulate and evaluate hypotheses concerning spatial distribution of diseases, mainly through prompt issuance of thematic maps (Barcellos & Sabroza, 2000).

Lau et al. (2012) argues that maps are particularly useful where disease surveillance and epidemiological data are sparse. According to Lau et al. (2012), predictive risk maps have been produced for many infectious diseases to identify risk areas for transmission and guide the allocation of public health resources.

The geographical information is relevant for public policy decisions, because it could be used for the planning of health services, development of early warning systems for outbreaks, allocation of disaster management resources, strengthening of infrastructure, and targeting interventions aimed at reducing infection risk as well as the overall burden of disease from leptospirosis (Lau et al., 2010).

Barcellos and Sabroza (2001) examined a leptospirosis outbreak that took place in the Rio de Janeiro Western Region (summer of 1996) using the Geographic Information System (GIS). Reis et al. (2008) used GIS to identify sources of leptospirosis in urban areas in a community-based survey of 3,171 slum residents in Salvador, Brazil. According to Reis et al. (2008) leptospirosis had become an urban health problem as slum settlements expanded leading to deficiencies in sanitation and other infrastructure. Confalonieri et al. (2011) created a vulnerability index for Rio de Janeiro State taking into account components of health, environmental and social characteristics, as well as the number of cases of infectious diseases influenced by climate.

The lowest index (associated to leptospirosis) occurred in the mid-South region of Rio de Janeiro state, while highest values were found in Costa Verde and Baixada Fluminense.

In Brazil there is no vaccine against leptospirosis in humans; the Ministry of Health, through the Secretariat of Health coordinates, advises and oversees surveillance and disease control that are developed throughout the country by the secretariats of states and local health.

Most data on the impact of the disease on SPMA dwellers is largely obtained from the clinic, hospital or mortality registry. The transmission of leptospirosis is usually associated with urine of rats and flood events during the summer (*rainy months*). But, it is important to higlight that this kind of data represent only one aspect of socio-environmental factors such as disease distribution and the risk of occurrence.

According to the Brazilian Health Surveillance Agency (2011), in São Paulo city 749 points of flooding were identified and the annual losses in the area reached nearly US$ 118 million. According to Young (2013), the proportion of surface at risk of flooding will be 23.5 % by 2030. The surveillance system included data from outbreaks associated with water. For this reason, state and local public health departments are primarily responsible for detecting and investigating waterborne-diseases. The analysis was on an outbreak, not an individual case.

For example, more than two persons must have experienced a similar illness after contact with or exposure to contaminated water in different settings (or settlements). Therefore, we associated the presence of slums, areas of boulevards and squares nearby rivers, channels and streams on flat terrains (up to 5% slope) with vulnerability to the disease. This was done in order to relate the frequency of floods (occurrences) and rats habitats (sewers). Coincidentally (or not), the flat areas around rivers and streams are occupied by the road system (avenues and streets with drainage and sewer channels).

Slums are considered vulnerable settlements because of the following characteristics: 1) inadequate access to safe water; 2) inadequate access to sanitation and urban infrastructure; 3) poor structural quality of housing and road system; 4) overcrowding; and 5) lack of public transport (United Nations, 2003).

According to IBGE (2010), the slum dwellers are economically, socially, and culturally excluded from the rest of the urban areas. Typically, the educational level and income of population is low; half of the population has no high school and undergraduate level, and receives less than one basic salary (US$ 250,00). The health conditions are also extremely deficient with precarious assistance (without health insurance or prevention).

Therefore, all of the population in the SPMA can be affected, but some areas are more vulnerable than others. The reason is the scathing exposure and lack of resilience in these areas. The health effects are expected to be severe for people living in flood areas lacking in sanitation, assistance or infrastructure.

However, according to Few et al. (2004), it is noteworthy that making the link between increased rainfall and flooding is not necessarily straightforward, because the flood outcome will also depend on the characteristics of the river basin.

The SPMA has a substantial number of *favelas* (shantytowns), industrial sites, rural and urbanized areas around the river banks, which facilitates the proliferation of leptospirosis because of the lack of sanitation. These are the same areas that suffer from flooding in the rainy months, increasing the likelihood of leptospirosis outbreaks. The floodwater invades buildings, squares, and roads close to rivers and exposes the population to contact with water (contaminated by rodent urine).

In São Paulo state, the time series from 1986 to 2011 confirmed 16,248 cases of leptospirosis, ranging from 239 to 964 cases annually. The incidence ranged from 0.84 to 2.34 cases per hundred thousand inhabitants. In the same period, more than 2,000 deaths were reported, ranging from 46.0 to 110 deaths (SINAN, 2012).

In the period 2007 to 2011, 4015 cases were confirmed and males (79.7%) were the most affected than females. Although there is no predisposition on gender or age for infection; there was higher risk among males in the age groups 35-49 years and 50-64 years than women. Leptospirosis occurred in many municipalities, with the highest incidences in the Capital (São Paulo city), where there were 1,281 cases (31,9%), and the São Paulo Metropolitan Area where 2,482 cases (61.81%) were recorded (SINAN, 2012).

It is believed the spread of leptospirosis is high among both the urban dwellers and health professionals. About 68% cases were recorded in urban areas, compared to 10% in rural areas, 4% in peripheral urban area. Among all confirmed cases, 45% occurred at home, 15% in workplace, 8% in leisure activities, 5% in other areas, whereas 27% were either ignored or not registered (SINAN, 2012).

With regard to incidence observed within 30 days prior to symptoms, in 40.8% were in the areas with high presence of rodents, 38.5% had contact with water or mud from floods, 25.9% had contact with river water or stream, lake or dam, 24.0% with garbage, 22.5% had direct contact with rodent urine, 17.7% with wasteland, 14.0% with sewage (SINAN, 2012).

According to municipal government of São Paulo (2012), about 1 831 477 people (16.43% of total population) live illegally on river banks and streams, and 33 435 (0.30%) live in dumps and landfills, and in both cases they live in São Paulo's slums. About 32.2% of slums suffer due to frequent floods particulalry during rainy seasons.

For these reasons, we considered some associated environmental factors, since the Tiete River Basin in which the SPMA is located consists of valleys prone to flooding (margins and riversides) that are highly urbanized. We used ArcGIS to integrate the spatial information and data on non-spatial attributes, where each spatial feature and its attribute information were linked. As a result, a map of vulnerability was produced based on the geographic risk appraisal.

The objective of the study was to evaluate leptospirosis at geographic locations based on environmental factors and produce a predictive disease risk map for the São Paulo Metropolitan Area (SPMA). Within this perspective, this map is a product (result) of integration between meteorological and public health information (such as precipitation, floods, hospital admissions, and cases of leptospirosis).

We evaluated the geographical areas that are most likely to be at risk of an increase in leptospirosis disease burden owing to the coexistence of the risk of climate change, environmental drivers of leptospirosis outbreaks and local socio-environmental circumstances.

METHODS

The SPMA is set in a sedimentary basin (Tiete River Basin) centred near 23°32′S″ and 46°38′W″, a low-lying region within the Atlantic Mountain Chain. It occupies 8,000 km^2 and it is surrounded by hills that vary from 650 to 1,200m in height and its proximity to the ocean influences its atmospheric circulation patterns. In addition, the SPMA comprises one of the largest human conglomerates in the world (19,956,590 people) with intense urbanisation according to the last census in the year 2010 (http://www. ibge.gov.br).

In this case study, we considered the previous analysis by Coelho and Massad (2012) of the leptospirosis hospital admissions (case numbers) and precipitation. The risk model estimated by Coelho and Massad (2012) showed that admissions due to leptospirosis have a significant correlation with season, peaking in the wet season (end of spring, summer and beginning of fall). From all the variables comprising the Brazilian Unique Health System data bank, Coelho and Massad (2012) selected the admission date, diagnosis, age and sex of the patients for analysis during a period of 7 years (daily data collection from January 1998 to December 2005). Subsequently, they also admitted "precipitation" as the main meteorological variable since this variable showed a significant variation in its pattern while temperature showed a homogeneous pattern for all lags.

Based on parameters and results obtained by Coelho and Massad (2012), we carried out an analysis in order to estimate the risk of leptospirosis infection during periods of precipitation and floods (Table 1).

In Figure 1 we show the relative risk (RR) of hospital admission as a function of flooding in steps of 20 mm of precipitation that proved significant. We continued the estimation up to 120 mm, following the same parameters observed by Coelho and Massad (2012).

Thence, a relative risk was considered in order to express the number of leptospirosis cases as a function of hospital admissions and flood events (in steps of 20 mm of precipitation). In this case, a flood ratio was calculated based on the highest precipitation value (120mm); when x = 1 (highest), the risk of the residents of SPMA contracting leptospirosis is 2.4%. Subsequently, we used ArcGIS to integrate the spatial information and non-spatial attribute data, where each spatial feature and its attribute information were linked.

Table 1. Parameters considered for development of the risk analysis

Precipitation (mm)	0	20	40	60	80	100	120
Increase hospital admissions (%)	0	15,6	33,9	55,0	79,7	108,5	142,1
RR of hospital admission	0	0,50	1,01	1,5	2,01	2,5	3,0
Flood (x)	0	0,2	0,3	0,5	0,7	0,8	1,0
Cases of leptospirosis	1006	1185	133	1549	1802	2089	2413

Source: Coelho and Massad, 2012.

Table 2. Qualitative criteria assigned to the areas susceptible to floods

Theme	Feature	Parte superior do formulário Flood Probability (criteria)Parte inferior do formulário
Hydrography	1. Major rivers	Very frequent (in lowland close to urban areas)
	2. Tributaries	Frequent (in lowland close to urban areas)
	3. Reservoirs	Not frequent
Land Use	1. Urbanized area	Very frequent (in lowlands close to rivers)
	2. Rural areas	Frequent (in lowlands close to rivers)
	3. Industry	Frequent (in lowlands close to rivers)
	4. Mines	Not frequent
	5. Slums	Very frequent (in lowlands close to rivers)
	6. Forestry	Not frequent
	7. Environmental Protection Areas	Not frequent
Slope	1. 0-5%	Very frequent
	2. 5-10%	Frequent
	3. 10-20%	Not frequent
	4. 20-30%	Not frequent
	5. 30-99999	Rare
Flooding points	1. Roads in raised areas and bridges	Not frequent
	2. Squares in raised areas	Not frequent
	3. Squares and tunnels in lowlands	Frequent
	4. Junction of roads in lowlands	Frequent
	5. Riverbanks (marginal tracks)	Very Frequent
Precipitation	1. 20,00mm	Small probability of flooding
	2. 40,00mm	Medium probability of flooding (cause = urbanization)
	3. 60,00mm	High probability of flooding (cause =
		urbanization)
	4. 80,00mm	High probability of flooding
	5. 100,00mm	High probability of flooding
	6. 120,00mm	High probability of flooding

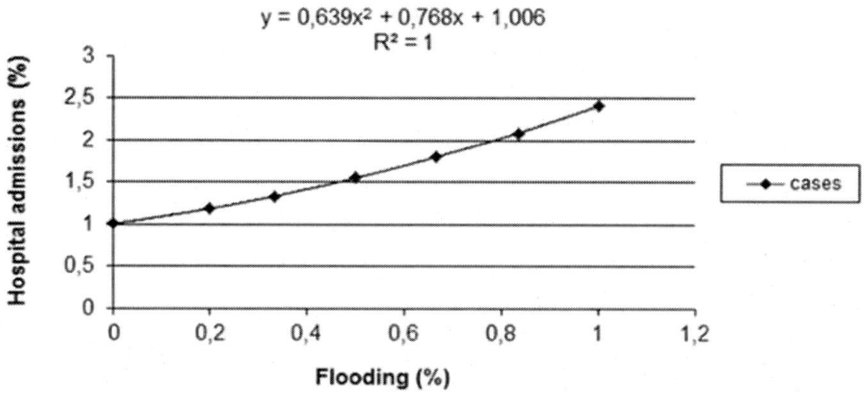

Figure 1. Cases of leptospirosis after flooding.

We established weights (based on the risk of the residents contracting leptospirosis) considering different sites (e.g., slums, urbanized, and rural areas, etc.) under different types of rain (e.g., 60mm/day, 80mm/day, etc.) and flood frequency (e.g., frequent, not frequent, very frequent, etc.).

We were able to arrange all the information spatially in a qualitative way by overlaying the spatial information layers, such as hydrography, land use, slope, flooding points, and precipitation (Table 2).

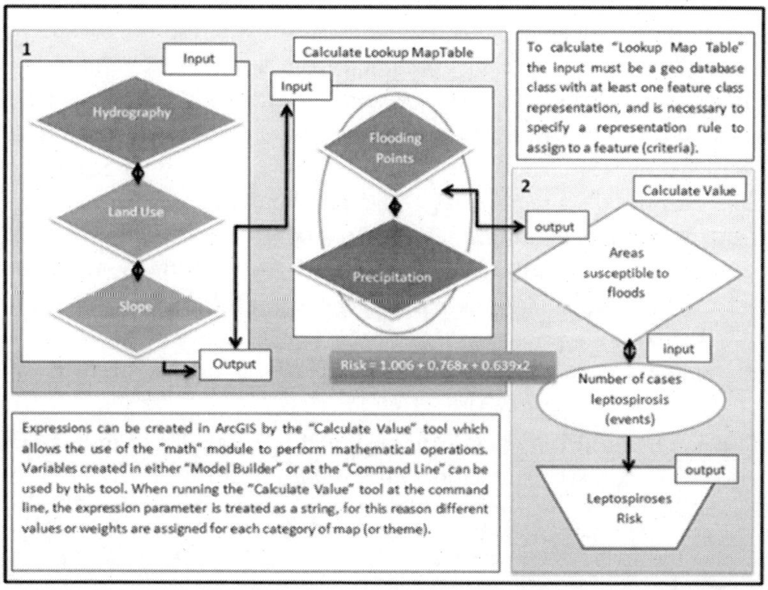

Figure 2. Analysis of themes (schematic flow).

In this case, environmental risk factors and triggers for outbreaks were considered in order to identify locations with high risk of leptospirosis infection, in other words areas most susceptible to a high disease burden.

These layers or thematic maps show a clear pattern of the distribution in each thematic case and indicate the core area of the metropolitan region. The ArcGIS also provided the "math" module to perform mathematical operations (Figure 2) in order to analyze the patterns and trends of the region.

Finally, sources of information about health (cases of leptospirosis) were processed through use of the "calculate value" tool, allowing epidemiological information to be available to support the spatial analysis.

RESULTS

The risk of contracting leptospirosis (or the probability that of an individual will develop leptospirosis) was calculated based on the number of flood events, according to the following equation:

$$\text{Risk} = 1.006 + 0.768x + 0.639x^2 \qquad (1)$$

where x = number of flood events

We used Microsoft Excel to calculate the polynomial equation automatically; therefore all the values in the equation were obtained through this process. In order to put in a spatial format and integrate the data from the equation (through the math module), it was necessary to consider the numerical values as qualitative information through specific criteria by assigning weights (Table 3). The results obtained from the qualitative analysis are presented in Table 4. They were central in terms of providing fundamental connections between empirical observation and mathematical expression of quantitative relationships.

Figure 3 below shows the areas most affected by flooding, which means that if there is a correlation between rainfall (precipitation) and number of leptospirosis cases, these could occur more frequently in areas of high risk located on the map. In this scenario, for every 20 mm precipitation, there is an average increase of 15.6% in the number of leptospirosis admissions (that is, the higher the precipitation the higher the admission rate). It should be noted that in the SPMA it can rain heavily (a peak of 120 mm was observed in a single day) in some periods (summer season), when leptospirosis can become a problem.

Table 3. The qualitative criteria assigned to intervals of leptospirosis and flood values

Risk of leptospirosis		Areas susceptible to floods		Classification
Intervals	weight	Intervals	weight	criteria (qualitative)
0-1	1	0-0,3	10	medium
1-2	2	0,3-0,7	20	high
2-3	3	0,7-1,0	30	very high
3-99999	none	1-9999	none	

Table 4. Criteria considered in qualitative analysis

Risk of Leptospirosis	Flood	Value Obtained	Criteria	Weight	Classification
1	10	11	Medium lepto + medium flood	1	Medium
2	10	12	High lepto + medium flood	2	High
3	10	13	Very high lepto + medium flood	2	High
1	20	21	Medium lepto + high flood	2	High
2	20	22	High lepto + high flood	3	Very high
3	20	23	Very high lepto + high flood	3	Very high
1	30	31	Medium lepto + very high flood	3	Very high
2	30	32	High lepto + very high flood	3	Very high
3	30	33	Very high lepto + very high flood	3	Very high

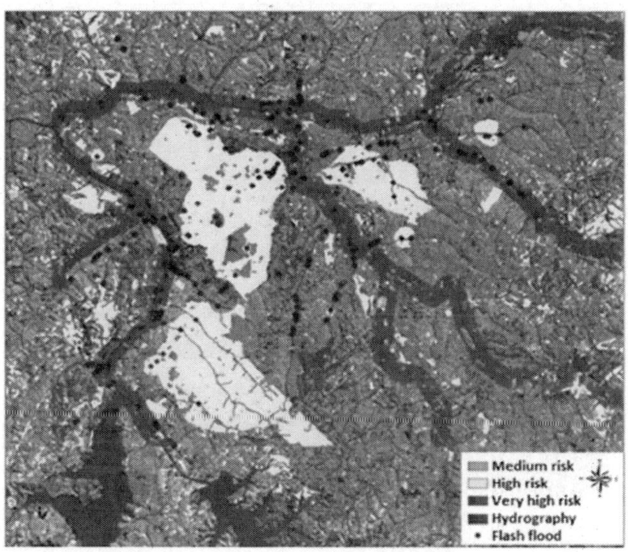

Figure 3. Estimation of areas susceptible to contracting leptospirosis in the SPMA.

DISCUSSION

Through this case study it was possible to develop a risk map by linking the information on rainfall and floods with the recorded number of cases of leptospirosis at hospital. This task was accomplished through integration of the data from the equation by assigning weights for different areas of the region. The results from the present investigation allowed identify probable areas at lepistopirosis risk. This study demonstrated the value of geographic information systems and disease mapping for identifying environmental risk factors for leptospirosis, and enhancing our understanding of its relation to climate.

We agree with Lau et al. (2012) that maps are particularly useful where disease surveillance and epidemiological data are sparse. According to Lau et al. (2012), predictive risk maps have been produced for many infectious diseases to identify risk areas for transmission and guide the allocation of public health resources. It is important to highlight that managing risk is an extremely difficult task, because it obviously depends upon context. The crucial logical step is identification of all initiating events and all possible sequences of events that may lead to serious consequences to public health. In the case of São Paulo Metropolitan Area, generally the survey time series data used to be sparse, making it difficult to carry out accurate analysis.

Due to the scale and the variables considered, it was not possible to identify the social stratification of the population and its relation to the cases of leptospirosis. We identified the areas most affected by floods and the relationship with hospitalization numbers, but we could not indicate where the most affected population was located at the time of contamination.

For example, the most infected men (or women) could be in their residences (e.g., in slums) or at work (e.g., in the center of the city), but it would be necessary to obtain more information, in different resolutions and broader scales, to identify this social stratification of space.

The geographical information is relevant for public policy decisions, because as mentioned by Lau et al. (2010) it could be used for the planning of health services, development of early warning systems for outbreaks, allocation of disaster management resources, strengthening of infrastructure, and targeting interventions aimed at reducing infection risk as well as the overall burden of disease from leptospirosis. In conclusion, the results from the present investigation made it possible to identify geographical areas in which the population shares an elevated risk of contamination. The identification of these areas can enable decisions to be made on strategies and interventions for improving the conditions of people in the São Paulo metropolitan area and similar areas around the world.

ACKNOWLEDGMENTS

This work was partially funded by the British Embassy in Brazil through the Strategic Programme Fund (SPF) from Foreign and Commonwealth Office - UK and National Institute of Science and Technology (Brazil) through the study "Vulnerability of Brazilian Megacities to Climate Change: Metropolitan Areas of São Paulo and Rio de Janeiro". It is important to highlight that this study could not be achieved without assistance from Micheline S. Z. S. Coelho.

REFERENCES

Barcellos, C., Sabroza P.C., (2000). Socio-environmental determinants of the leptospirosis outbreak of 1996 in western Rio de Janeiro: a geographical approach. *International Journal of Environmental Health Research,* 10(4): 301-13.

Barcellos, C., Sabroza P.C., (2001). The place behind the case: leptospirosis risks and associated environmental conditions in a flood-related outbreak in Rio de Janeiro. Cad Saude Publica, Suppl 17: 59-67.

Bharti, A.R., Nally J.E., Ricaldi, J.N., Matthias MA, Diaz M.M., Lovett, M.A., Levett, P.N., Gilman RH, Willig MR, Gotuzzo E, Vinetz JM (2003) Leptospirosis: a zoonotic disease of global importance. *Lancet Infect Dis.,* 3: 757–771.

Biosci, J.,Vijayachari, P., Sugunan, A.P., Shriram, A.N. (2008) Leptospirosis: an emerging global public health problem. Nov; 33(4): 557-69.US National Library of Medicine. National Institutes of Health http://www.ncbi.nlm.nih.gov/pubmed/19208981.

Coelho, M.S.Z.S. Massad, E. (2012). The impact of climate on leptospirosis in São Paulo, Brazil. *International Journal of Biometeorology,* 56: 233–241.

Confalonieri, U., Barata, M., de Lima, A.C.L., Marinho, D. P., Luigi, G., De Simone, G. C., de B Ferreira, I., Pinto, I.V. (2011) Mapa de Vulnerabilidade da População do Estado do Rio de Janeiro aos Impactos das Mudanças Climáticas nas Áreas Social, Saúde e Ambiental. Fundação Fiocruz. Rio de Janeiro.

Farr, R.W. (1995). Leptospirosis. ClinInfectDis 21:1–6.

Field C.B., Barros V., Stocker T.F., Qin D., Dokken D.J., Ebi K.L., Mastrandrea M.D., Mach, K.J., Plattner, G.K., Allen, S.K., Tignor. M., Midgley, P.M., (2012) Managing the Risks of Extreme Events and Disasters to Advance Climate Change Adaptation. A Special Report of Working Groups I and II of the Inter-governmental Panel on Climate Change (IPCC). Cambridge University Press, Cambridge, UK, and New York, NY, USA, pp582.

Few, R., Ahern, M., Matthies, F., Kovats, S., (2004) Floods, health and climate change: A strategic review. Tyndall Centre for Climate Change Research. Working Paper No. 63,: 138.

IPCC, (2007). Climate Change 2007: Synthesis Report. Contribution of Working Groups I, II and III to the Fourth Assessment Report of the Intergovernmental Panel on Climate Change IPCC, Geneva, Switzerland, 104 pp.

Lau, C.L., Clements, A.C., Skelly, C., Dobson, A.J., Smythe, L.D., Weinstein, P. (2012) Leptospirosis in American Samoa – Estimating and Mapping Risk Using Environmental PLoSNegl Trop Dis. 2012 6(5): 1669.

Lau, C.L., Smytheb, L.D., Craigb, S.B., Weinsteina, P., (2010) Climate change, flooding, urbanisation and leptospirosis: fuelling the fire? Transactions of the Royal Society of Tropical Medicine and Hygiene, 104: 631–638.

Levett, P. N., (2001) Leptospirosis. ClinMicrobiol Rev. 14.:296–326.

McBride, A.J., Athanazio, D.A., Reis, M.G., Ko, A.I., (2005) Leptospirosis. *Curr. Opin. Infect Dis.,* 18: 376–386.

Nobre, C.A.; Young, A.F.,Saldiva, P.H., Marengo, J.A., Nobre, A.D.,Ogura, A., Thomaz O., Parraga, G.O.O., Silva, G.C.M., da Valverde, M,, Silva, G., Silveira, A.C, Rodrigues,G., de O,., (2011). Vulnerabilidade das Megacidades Brasileiras às Mudanças Climáticas: Região Metropolitana de São Paulo. Relatório Final. Eds. Nobre, C A e Young, A.F.,. INPE, UNICAMP. São Paulo, Brasil, 178.

Reis, R.B., Ribeiro, G.S., Felzemburgh, R.D., Santana, F.S., Mohr, S., Melendez, A.X., Queiroz, A., Santos, A. C, Ravines, R.R, Tassinari, W. S., Carvalho, M..S, Reis, M.G. and Ko, A.I. (2008) Impact of environment and social gradient on leptospira infection in urbanslums. *PLoSNegl Trop. Dis.,* 2(4): 228.

Soares, T.S., Latorre, M., R., Laporta, G.Z., Buzzar, M.R., (2010) Spatial and seasonal analysis on leptospirosis in the municipality of Sao Paulo, Southeastern Brazil, 1998 to 2006. *Revista de saude publica,* 44(2):283-91.

Vanasco, N.B., Kemerer, R., Oliva, M.E., (2004) Brote de leptospirosis rural en un tambo de la provincia de Entre Ríos, Argentina, febrero-marzo 2003. Salud (i) Ciencia 12: 26–31.

In: Flooding
Editors: M. R. Motsholapheko and D. L. Kgathi
ISBN: 978-1-63117-607-4

Chapter 5

CHANGES AND THREATS OF THE TAIPEI FLOOD PREVENTION SYSTEM

***Shang-Shu Shih*[1,•]*, Jihn-Sung Lai*[1] *and Gwo-Wen Hwang*[1]**
[1]Hydrotech Research Institute, National Taiwan University,
Taipei City, Taiwan ROC

ABSTRACT

Due to dense population living along the riverside, river flooding is one of the major hazards in Taiwan. The Taipei metropolitan area is the largest city with more than one-third of the total population of Taiwan. In this area, numerous severe flooding disasters caused by typhoon events have occurred and resulted in heavy losses. To mitigate flood-related disasters, a large-scale flood prevention program was implemented in 1963 and fully completed in 1999, namely the Taipei Flood Prevention System. The specific goal was to protect the Taipei metropolitan area against the 200-year recurrence flood in the Tanshui River. Levees and dykes of 32 km in length were constructed and improved along the river. Mitigation of floods for the Tanshui River system is inhibited by the bottleneck, which occurs at the smallest river width near the Taipei Bridge. Therefore, the Erchung Floodway was established to divert some of the flood water. The specific goal of the Erchung Floodway was to divert 9200 m^3/sec peak flood discharge under a 200-year return period flood. However Erchung Floodway's function has been changed over time by urban development and natural alterations including river sand mining, riverine park construction, riverine plant succession, bridge construction and so on. Riverbed elevation changes due to sand mining are believed to have influenced the strength of the Taipei Flood Prevention System and thus were discussed in this study. The current protection criteria and impacts from anthropogenic effects and climate change threats were also examined. Both the physical and numerical models were used and analyzed. The sensitivity analysis of thirty-two scenarios corresponding to four factors has been investigated, including riverbed elevation, riverbed roughness, and water stage at the river mouth under the Q_{200} flood. The simulated results show that the flood diversion capacity of the Erchung Floodway, a key infrastructure for dividing floods in the Taipei Flood Prevention System, has decreased by 30%. We also found that the Taipei Flood

• Phone: +886-2-33662624; E-mail: uptreeshih@ntu.edu.tw.

Prevention System will encounter challenges if the riverbed roughness in the Erchung Floodway increases by over 50%, the riverbed roughness in the Tanshui River increases by over 25%, and the Q_{200} increases by over 13%. We conclude that the degrading process of the Taipei Flood Prevention System due to rapid urbanization and the corresponding strategies including river roughness and riverbed elevation control are meaningful lessons especially for developing countries. A comprehensive and effective evacuation program and monitoring system is also suggested.

Keywords: Taipei Flood Prevention, Tanshui River, climate change, anthropogenic effects, floodway

INTRODUCTION

Population growth and improved living standards along with the possession of high-value belongings have created new challenges and have increased the vulnerability of society and environment towards flood hazard (Dang et al. 2011). Due to the rapid urbanization of Taiwan, approximately 80% of the population lives in urban areas (Huang and Hsu 2003). These urban areas are usually near large rivers due to their inherent industrial and domestic water demands. Flooding and inundation along riversides caused by the heavy precipitation that is associated with rain storms or typhoons frequently occur in lowlands and floodplains (Chen et al. 2006; Hsieh et al. 2006; Pan et al. 2012). Therefore, river floods have become a major cause of concern in Taiwan. Understanding the impact of extreme hydrological events is becoming essential from both scientific and political perspectives (Lehner et al. 2006).

Although Taiwan has used potential inundation maps as references to set up non-structural strategies for mitigating flood hazards (Murphy 2003; Lowe 2003; Chen et al. 2006), structural measures still play an important role in decreasing the risk of flooding. Three protection measures, including the construction of high-level protection levees, diversion channels, and detention reservoirs, have been proposed for flood mitigation (Hsieh et al., 2006). To mitigate flood disasters a mega-project, Taipei Flood Prevention System, was implemented in 1963 and completed in 1999. The specific goal was to protect the Taipei metropolitan area against the 200-year return flood period (Ministry of Economic Affairs, 1973). Approximately, a total of 32 km of dykes and levees have been constructed (Huang and Hsu, 2003). However, its function may have changed over time due to urban development, in the form of hydraulic facility construction, and natural alterations resulting from riparian vegetation succession and sediment deposition. Furthermore, climate change effects might degrade the flood prevention function of the Taipei Flood Prevention System and increase the flood risk of the Tanshui River. Dang et al. (2011) indicated that climate changes are projected to increase the frequency and severity of extreme weather events. An observation in Central Europe shows that flood risk and vulnerability are likely to increase due to climate change (Kundzewicz et al. 2005). IPCC (2001) concluded that consequent adaptation strategies are crucial for preserving the health of human societies and a sustainable environment, because the intensity and frequency of floods may be raised due to climate change. Consequently, economic losses caused by natural catastrophes could increase significantly (Botzen et al. 2010).

The Tanshui River system passes through the Taipei metropolitan area where more than one-third of the Taiwanese population lives. It consists of three main tributaries: the Dahan Creek, the Hsindian Creek, and the Keelung River, with a mainstream length of 158 km and a watershed area of 2,776 km^2 (Figures 5-1 and 5-2). The Taipei metropolitan area encountered numerous severe floods from the Tanshui River system, and these resulted in heavy losses. The main challenge in the Tanshui River system is that a bottleneck occurred at the smallest river point (which had a width of 450 m) near the Taipei Bridge (T024A). This river point could not allow water from a 200-year return period flood to pass. The Erchung Floodway was thus established to divert some of the floodwater. It is 450 m to 650 m in width and uses a fixed weir as an inlet control. The weir crest is EL. 4.0 m, the external part of the weir is EL. 3.0 m, and the internal part of the weir is EL. 2.5 m (Water Resources Planning Commission MOEA, 1980; Water Resources Planning Commission MOEA, 1983). The specific goal for the construction of the Erchung Floodway was to arrive at 9,200 m^3/sec division under the 200-year flood event. The Q_{200}-Tanshui River could thus be decreased from 23,500 to 14,300 m^3/sec (Water Resources Planning Commission MOEA, 1980).

Even though projects such as the Taipei Flood Prevention System are effective in mitigating flood disasters, they require considerable time and money to implement. In addition, the function of flood prevention might be changed over time by urban development and it may be difficult to detect this change before a catastrophic flood occurs. Therefore, this chapter aims to examine anthropogenic effects on the Taipei Flood Prevention System using physical model experiments and numerical model simulations. Threats of climate change such as a rise in sea level and flood discharge are also discussed.

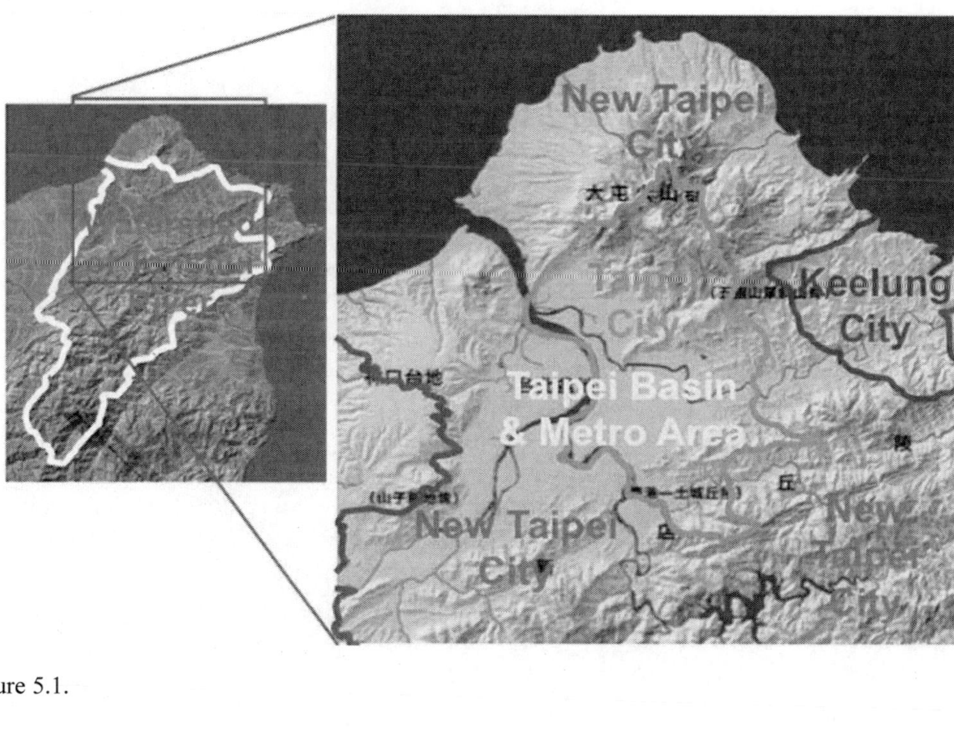

Figure 5.1.

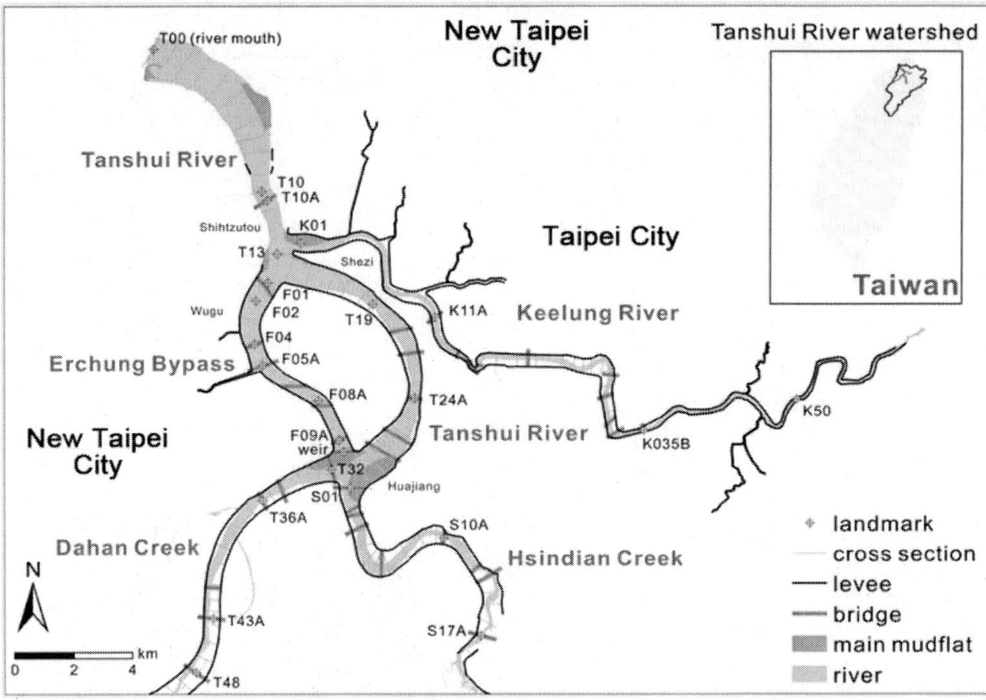

Figure 5.2.

Taipei Flood Prevention System

Historical Overview

The construction of the river bank and development of levees on the Tanshui River started in 1898 during the period of Japanese occupation. After the Taiwan retrocession, the Taiwan government authority promoted local flood controls to protect the residents and properties of Taipei city. However, due to inadequate and insufficient facilities, they were unable to prevent large flood events, therefore the flood disasters occurred frequently. In order to solve the inundation problem along the Tanshui River, the government started a flood prevention project in 1963. This project included the construction of dykes, pumping stations, and diversion channels. The first phase of the project was completed in 1965. The project was revised in 1973 and re-implemented in 1979. The modified project was completed in 1999.

The Taipei metropolitan area, being the largest in Taiwan and including more than one-third of the total population, deserves special attention with regard to flood prevention measures. In the past, numerous severe floods with heavy losses have occurred in the Taipei metropolitan area, such as the floods caused by Typhoon Pamela in 1961, Typhoon Opal and Typhoon Amy in 1962. To mitigate such flood disasters, a large-scale flood prevention program was implemented in 1963 and fully completed in 1999, namely the Taipei Flood Prevention System. Since 1981, dykes and levees, approximately 32 km in length, have been constructed along the major rivers in this region (Huang and Hsu 2003). The specific goal of this measure was to protect the Taipei metropolitan area against the 200-year flood (Figure 5-3).

In summary, the history of the construction of flood controls on the Tanshui River can be divided into the following four stages: a) the period of Japanese occupation (1898-1949), b) temporary flood control projects (1949-1963), c) flood control along the Tanshui River (1963-1965), and d) the Taipei Integrated Flood Mitigation Project (1973-1999).

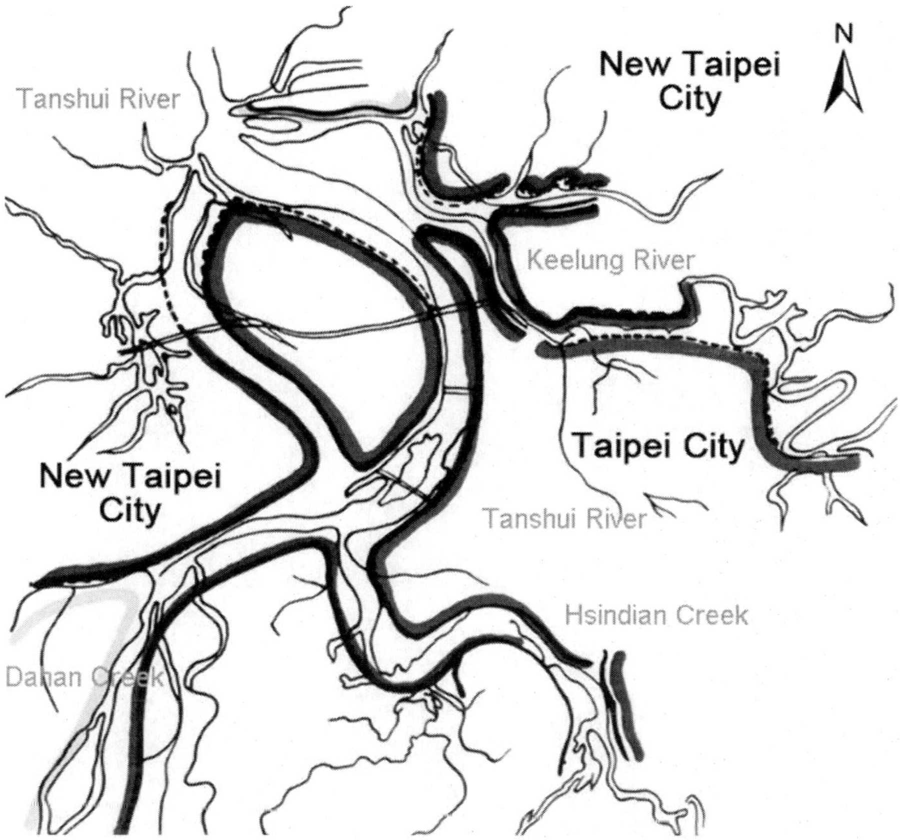

Figure 5.3.

Erchung Floodway Diversion

The Erchung Floodway is located in an area settled under the reclamation policy of the Qing Dynasty. Various grains were cultivated and paddy fields were built for planting rice. The plentiful supply of food allowed many people to settle there. Thus settlements, such as Joutzuwei, were formed. This area was an important agricultural area in Taipei until the Japanese occupation. However, in 1964, the government dredged the Tanshui River and broadened the river width of Guandu, which was the narrowest section of Tanshui River. As a result, tides began to invade Wugu, Shinjuang and other lower lands. Because the water did not subside gradually, a vast swamp was formed. The Erchung Floodway, one of the successful measures of the Taipei Flood Prevention System, was established between 1982 and 1996 to divide the Tanshui River floodwaters and to protect the residents living in the Taipei metropolitan area.

The specific goal of the Erchung Floodway diversion was to divide 9,200 m^3/sec of floodwater from the Tanshui River under a 200-year recurrence flood event. This would allow the Q200-Tanshui River to decrease from 23,500 m^3/sec to 14,300 m^3/sec. Since its completion, the Erchung Floodway diversion has functioned successfully during ten flood events as listed in Table 5-1 below. Over time, the environment changed due to industrial development and natural alternations (Figure 5-4).

Figure 5.4a.

Figure 5-4b.

Table 5.1. Historical water level recordings of the diversion channel stations

Typhoon events	Highest water level (m)	Time of occurrence
Herb	5.27	1996/08/01 03:00
Zeb	5.16	1998/10/16 19:00
Xangsane	4.83	2000/11/01 14:00
Nari	4.91	2001/09/17 03:00
Aere	6.62	2004/08/25 08:00
Matsa	4.86	2005/08/05 13:00
Krosa	6.15	2007/10/06 20:00
Sinlaku	4.79	2008/09/13 23:00
Jangmi	5.13	2008/09/28 21:00
Sola	6.55	2012/08/02 08:00

RIVER SAND MINING AND CORRESPONDING RIVER MORPHOLOGICAL CHANGES OF THE TANSHUI RIVER SYSTEM

We measured and calculated temporal and spatial distribution of erosion and deposition changes on the river network of the Tanshui River system by using a cross-section field data provided by the 10th River Management Office, Water Resources Agency. The first survey of cross-sectional bathymetry began in 1969. Sand mining was banned on some streams in 1982, and seven years later sand mining was banned on the whole river. The third phase of the Taipei flood implementation plan was completed in 1999. From 2002-2005, the Yuanshanzih flood diversion was built. Therefore, we divided the riverbed changes of the Tanshui River system into five periods: 1969-1981 1982-1989 1990-2000 2001-2012 1990-2012.

Furthermore, based on tidal water levels in the inter-tidal zone boundary, we divided the whole river system into the following eight river sections: Tanshui River (1), Tanshui River (2), Dahan Creek (3), Dahan Creek (4), Hsindian Creek (5), Hsindian Creek (6), Keelung River (7), and Keelung River (8). The Keelung River (8) has no cross-section data before 2001.

The riverbed slope results exhibit that the slope of the tidal region is significantly less than that of the non-tidal region, i.e., the tidal region slope is less than 1/1000, while the slope of the non-tidal region is greater than 1/1000. The non- tidal section of the riverbed slope can be sorted by size as follows: Dahan Creek (4) > Keelung River (8) > Hsindian Creek (6). Only the riverbed slope of the tidal section near the estuary of the Tanshui River (1) is an inverse slope. Figure 5-5 shows the longitudinal elevation analysis results. The effect of tides on the riverbed is more significant than any other factor. The historical amount of slope change on the Tanshui River (1) is the smallest compared with the other rivers. The Dahan River (3) reached a peak slope in 1982 and the Hsindian Creek (5) peaked in 1989. Since then, the slopes in these sections have been decreasing. After 2001, the slopes of Dahan Creek (3) and Hsindian Creek (5) are similar and the slope of the Keelung River (7) is relatively moderate, with values close to that of the Tanshui River (2). The slopes of the four rivers have stabilized. (Table 5-2)

Table 5.2. The riverbed slope of each river reach of the Tanshui River system in different years

River reach	Slope ×1000				
	1969	1982	1989	2001	2012
Tanshui River (1)	-0.262	-0.339	-0.228	-0.466	-0.101
Tanshui River (2)	0.199	0.313	0.358	0.194	0.163
Dahan Creek (3)	0.765	1.142	0.865	0.529	0.435
Dahan Creek (4)	4.241	4.299	4.282	4.278	4.206
Hsindian Creek (5)	0.443	0.414	0.776	0.629	0.476
Hsindian Creek (6)	2.027	1.819	1.836	1.573	1.450
Keelung River (7)	-	-	-	0.205	0.196
Keelung River (8)	-	-	-	2.640	2.649

-: no data

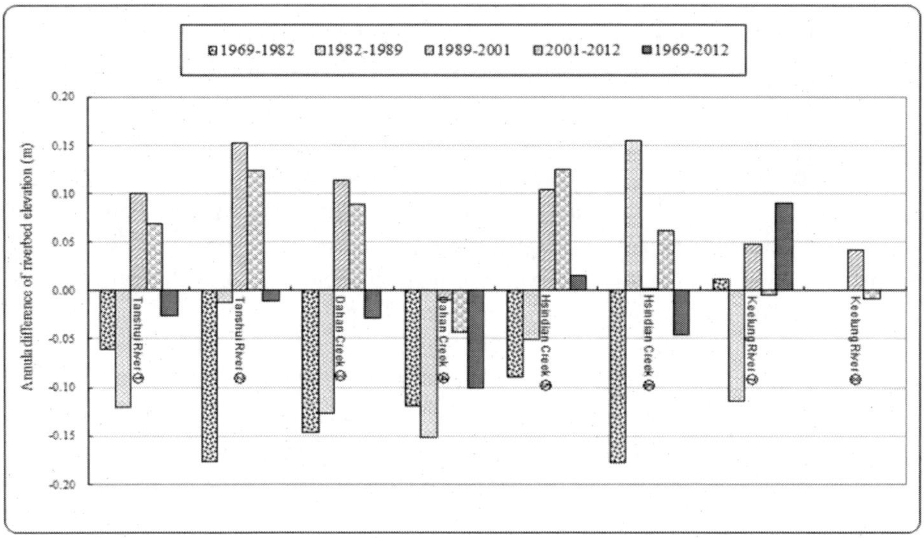

Figure 5.5.

Figure 5-5 illustrates the annual differences in riverbed change in different time periods. Analysis of annual riverbed elevation changes from 1990 to 2012 shows the deposition thickness of the tidal section was significantly greater than that of the non-tidal section. The total average riverbed deposition thicknesses of the tidal sections of Tanshui River (1), Tanshui River (2), Dahan Creek (3), Hsindian Creek (5), and Keelung River (7), were 1.19 m, 1.95 m, 1.47 m, 1.47 m, and 0.61 m, respectively. The total slope of the Tanshui River (2) increased significantly at the riverbed, while the riverbed of the Keelung River (7) showed a relatively moderate increase. In addition, Figure 5-5 also represents that the tidal section of the Tanshui River has exhibited continued siltation after the sand mining ban (1989) but the sedimentation rate has slowed.

The spatial river erosion occurring between 1969 and 1982 (Figure 5-6) shows that sand mining had the noticeable effect of lowering the riverbed. From 1982 to 1989, except for the Hsindian Creek (5), the rest of the river still had significant river erosion (Figure 5-7). From

1990 to 2012, except for the Dahan Creek (4), the remaining river siltation data revealed that after the total ban on sand mining, the riverbed has shown a gradual siltation trend (Figure 5-8).

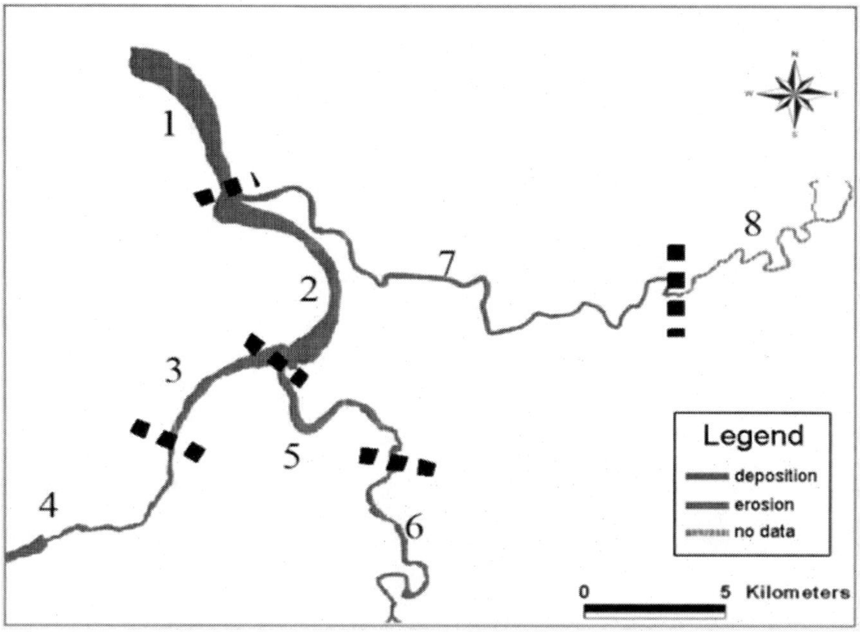

Figure 5.6.

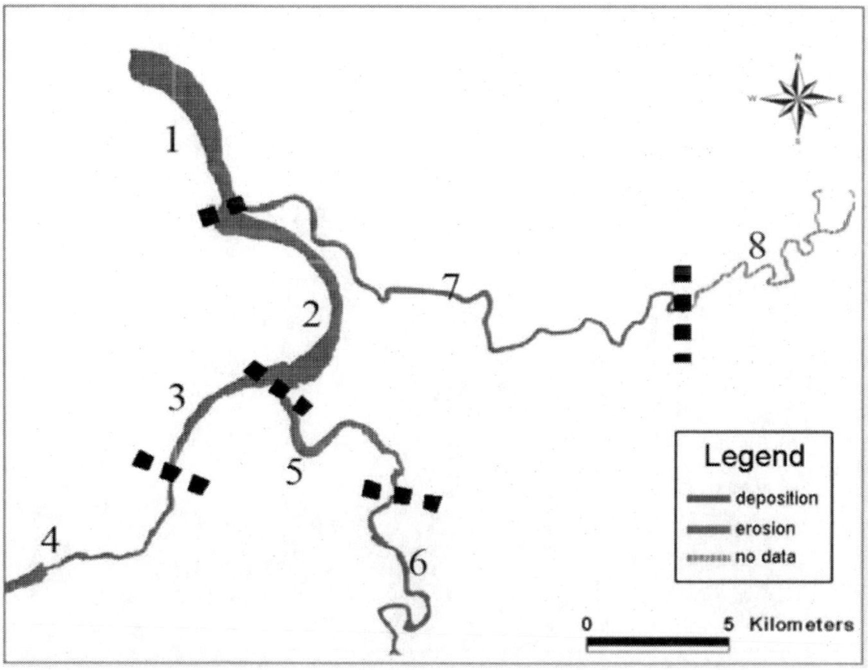

Figure 5.7.

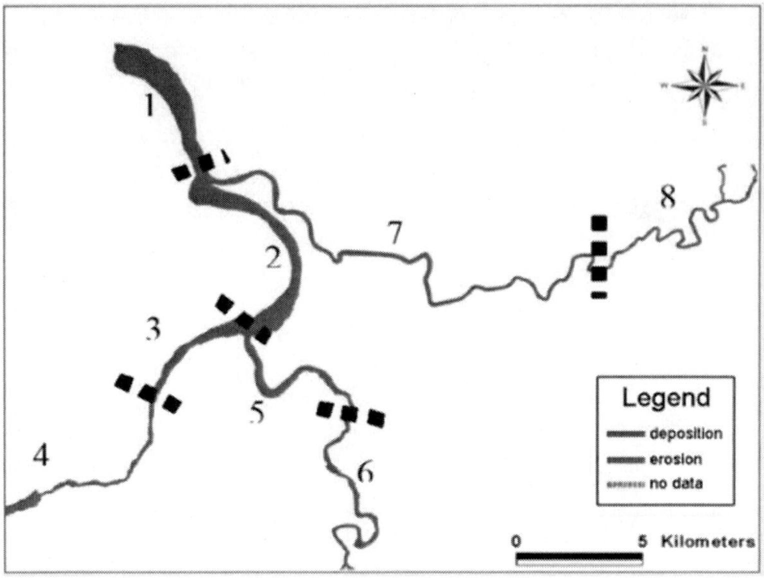

Figure 5.8.

METHODS

Physical Model Experiments

The physical model used in this study was the distorted and rigid bed model: the vertical scale was 50 (prototype/model) and the horizontal scale was 300 (prototype/model) (Figure 5-9).

Figure 5.9.

The Froude Number of the model was identical to the prototype. This model was built in 1963 and reconditioned in 1972, 1976, 1985, 1996, 1997, 2000, 2002, 2003, 2004 and 2011 to meet the changes of bathymetry, levees, or bridges. To analyze flooding in the current Taipei Flood Prevention System and for numerical model verification, the physical model experimental discharges were 11,500 m^3/sec and 13,200 m^3/sec at T43A for the Dahan Creek, 10,200 m^3/sec and 10,300 m^3/sec at S17A for the Hsindian Creek, and 1,300 m^3/sec and 1500 m^3/sec at K35B for the Keelung River under the 100-year and 200-year return period floods (Water Resources Agency, Ministry of Economic Affairs, 1996). The downstream boundaries were 2.3 m and 4.03 m at the river mouth (T00), but the model boundary was at 4.5 km offshore from the river mouth.

Numerical Model Simulations

Due to the submerged flow type of the Erchung Floodway, the weir formula could not be used to estimate bypass discharge. For example, the water stage at the weir increased by 0.13 m whereas the Q_{200}-Erchung Floodway decreased from 8,225 to 7,755 m^3/sec when comparing Report 1980 and Report 1983. Therefore, the one-dimensional model, HEC-RAS (the Hydrologic Engineering Centers River Analysis System) was used in this study as a relatively cheap tool to estimate bypass discharge and several numerical experiments. In addition to current threats, anthropogenic effects and climate change threats will lead to severe challenges for flood prevention in Taipei. The Taipei Flood Prevention System is complicated. Therefore potential vulnerabilities and an appropriate strategy to maintain or restore flood prevention could be identified through the numerical model simulation for sensitivity analysis of influential factors. The HEC-RAS model has been successfully applied to networking rivers under subcritical, supercritical or mixed flow conditions, and is able to automatically estimate bypass discharges for junctions based on the energy or momentum equilibrium method (Hydrologic Engineering Center, 2010[a], 2010[b]). It was also adopted in the cross-section bathymetry of the Tanshui River system in 2009 as topographic data: the upper boundary is at T48 for the Dahan Creek, at S017A for the Hsindian Creek and at K050 for the Keelung River. All bridges within this study area were contained in the model. Orthographic projection was adopted to consider the crossing areas where the directions of the bridges are diagonal with the main stream direction.

Sensitivity Analysis

Investigation of Influential Factors

One way to determine the influential factors of Taipei Flood Prevention System is through comparison of experimental results of previous physical models. All experiments adopted the same upstream boundary condition, Q_{200} (the values of discharge are 13,200 m^3/sec for the Dahan Creek, 10,300 m^3/sec for the Hsindian Creek, and 1500 m^3/sec for the Keelung River), but different downstream boundary condition. There were two different water stages at the river mouth (T00), EL. 2.30 m and EL. 4.03 m. In order to compare with these former experimental results, both water stages were used as the downstream boundary condition in our physical model experiments.

Comparing Report 1974 and Report 1980, the Q_{200}-Tanshui River increased by about 1,000 m^3/sec while the Q_{200}-Erchung Floodway decreased from 9,200 m^3/sec to 8,225 m^3/sec (Table 5.3). However, the water stage at T024A (the Taipei Bridge), the bottleneck of the Tanshui River, rose by a mere 5 cm due to riverbed erosion, which resulted from sand mining activity.

Table 5.3. Comparison between different physical model experiments results

Report [a]	1974	1980	1983	1996	2011[a]	2011[b]
1. Bathymetry year	1970	1976	1976	1994	2010	2010
2. Water stage at T00 (m)	4.03	--	--	2.30	2.30	4.03
3. Number of bridges						
In Erchung Floodway	2	3	3	3	7	7
In Tanshui River	6	6	9	22	23	23
4. Q_{200} (m^3/sec)						
In Erchung Floodway	9,200	8,225	7,755	6,500	6,300	6,500
In Tanshui River at Taipei Br.	14,300	15,275	15,745	17,000	17,200	17,000
5. Water stage (m)						
At T024	8.40	8.45	8.47	8.34	8.11	8.38
At Erchung Floodway weir	9.61	9.17	9.30	9.20	9.10	9.44
6. Mean riverbed elevation (m)[c]						
At T031	0.8	-0.4	-0.4	-1.3	-0.2	-0.2
At T024A	-3.0	-3.6	-3.6	-5.0	-1.5	-1.5
At T010A	-2.1	-2.9	-2.9	-4.2	-4.0	-4.0

[a] Report 1974 referred to the Report of the Tanshui River physical model experiments (1966), Review for Taipei Flood Prevention (1970), Report on the work of the technology group of Taipei Flood Prevention 03-(Flood)-11 (1972), Proposal for Taipei Flood Prevention (Draft)(1973), and the Experimental report for arrangement and stability of the Tanshui River Bypass entry (1974). Report 1980 referred to the Report on the physical model experiments of the Erchung Floodway entry（1980; Report 1983 referred to the report on Related projects to preliminary Taipei Flood Prevention: Presentation on the physical model experiments of the Erchung Floodway（1983）; Report 1996 referred to the Experimental report for the physical model of the Tanshui River establishment and Taipei Flood Prevention performance（1996）; Report 2011a and Report 2011b referred to the experimental results of this study, which adopted 2.30 m and 4.03 m as the water stage at T00.

[b] Riverbed bathymetry used the 2010 data provided by the Tenth River Management Office, Water Resources Agency, but the inlet and outlet of the Erchung Floodway were further modified based on the latest and detailed measurement results in 2011.

[c] The values were calculated from the data of riverbed bathymetry.

The activity of sand mining was permitted in the past and banned in 1989. After 1989, sediments kept silting in the Tanshui River and raised the riverbed. For instance, the mean elevation of the riverbed at T024A declined by 0.6 m in the period 1970 to 1976, but significantly rose to more than 3.5 m in the period 1994 to 2010. Due to low riverbed elevation, the Tanshui River is able to convey more flood discharge. The riverbed elevation is thus one of the key influential factors on Taipei Flood Prevention. It is conducive to relieving flood pressure, even though the bypass discharge decreases. Comparing Report 1974 and Report 2011[b], the Q_{200}-Erchung Floodway decreased from 143,000 m^3/sec to 17,000 m^3/sec, but the water stage at T024A was almost the same.

The Report 1980 and Report 1983 both adopted the same bathymetry and hence the impact of bathymetry could be eliminated. But at the Q_{200}-Tanshui River increased from 15,275 m^3/sec to 15,745 m^3/sec. The major difference between these two experiments was due to the construction of the Guandu Bridge (T010A) in the Tanshui River. This bridge was built at another bottleneck of the Tanshui River and it induced backwater effect. The backwater effect raised water stages of the Tanshui River and the Erchung Floodway and reduced bypass discharge. On the other hand, comparing Report 1974 and Report 1996, the Q_{200}-Tanshui River increased from 14,300 m^3/sec to 17,000 m^3/sec. The water stage at T024A still declined from 8.40 m to 8.34 m, because the water stage at T000 declined from 4.03 m to 2.30 m at the same tidal condition (the tide stage at 4.5 km outside the river mouth was 1.91 m). The difference between the water stages was caused by a wharf construction and river mouth shape alteration. Furthermore, Report 2010[(a)] and Report 2010[(b)] adopted the same bathymetry but different water stages at T000. The experimental results showed that the Q_{200}-Erchung Floodway increased to 200 m^3/sec and the water stage at T024A increased from 8.11 to 8.38 m whereas the water stage at T000 increased from 2.30 m to 4.03 m. Therefore, the water stage at the river mouth is another key influential factor for Taipei Flood Prevention.

The riverbed elevation (anthropogenic effect) and the water stage (climate threat) at the river mouth were the key influential factors on the Taipei Flood Prevention. However, two key influential factors have not been regarded in these experiments: a) roughness of riverbed (anthropogenic effect) and b) the 200-year return flood period (climate threat).

Scenarios Set up for Sensitivity Analysis

A systematic sensitivity analysis from numerical model experiments is an effective way to investigate the impacts of environmental factors on the abilities of flood prevention measures.

The variations in riverbed elevation and riverbed roughness were selected as two key factors that impact on the ability of the Taipei Flood Prevention System. The climate change threats also lead to severe challenges for the Taipei Flood Prevention System. Two factors were selected to be examined: a) the water stage at the river mouth and b) the flow discharge of the 200-year recurrence flood.

Thirty-two scenarios were examined (Table 5-4) in three stages using the variation of the aforementioned key impact factors. First, the riverbed alterations in the Erchung Floodway diversion and in the Tanshui River were examined; both of these structures influence the strength of the Taipei Flood Prevention System. The topography in our numerical model was further modified to measure the impact of riverbed alteration. Second, a parameter in the numerical model, Manning's n value, was modified to analyze the impact of riverbed roughness on the strength of the Taipei Flood Prevention System.

Table 5.4. Scenario settings for systematic sensitivity analysis

Impact factors	Scenarios
Anthropogenic effects	
Riverbed elevation	Modified the topography data in the Erchung Floodway diversion (whole) Year 1986 Year 2010 (current situation) Modified the topography data in the Erchung Floodway diversion (divide into upstream, midstream, and downstream) Year 1986 Year 2010 (current situation) Modified the topography data in the Tanshui River (Taipei Br.) Year 1970 Year 1986 Year 2010 (current situation)
Riverbed roughness	Modified Manning's N values in the Erchung Floodway diversion (whole) -50% -25% +0% (current situation) +25% +50% +100% Modified Manning's N values in the Erchung Floodway diversion (divide into upstream, midstream, and downstream) -50% -25% +0% (current situation) +25% +50% +100% Modified Manning's N values in the Tanshui River (Taipei Br.) -50% -25% +0% (current situation) +25% +50% +100%
Climate change threats	
Water stage at river mouth	Modified the downstream boundary condition (water stage) 2.30 m (current situation) 2.90 m 3.50 m 4.03 m 5.03 m
200-year recurrence flood	Modified the upstream boundary condition (flow) 25,000 m^3/sec (current situation) 28,300 m^3/sec

The values of Manning's n in the Erchung Floodway diversion and in the Tanshui River were further modified to -50%, -25%, +25%, +50% and +100%. Third, five scenarios were set to analyze the impact of water stage at the river mouth (downstream condition of the Taipei Flood Prevention System): 2.3, 2.9, 3.5, 4.03, and 5.03 m at T00. Considering extreme rainfall due to climate change, the 200-year recurrence flood (the total discharge of the Dahan Creek and the Hsindian Creek), that is the upper boundary condition of the Taipei Flood Prevention System, increased from 25,000 m^3/sec to 28,300 m^3/sec (Water Resources Planning Institute WRA 2010).

RESULTS

Physical Model Validation

Three typhoon events, in which the floods were divided by the Erchung Floodway, were used for physical model validation. The peak flows of Typhoon Jangmi were used to calibrate riverbed and flood plain roughness of the model, and peak flows of Typhoon Korosa and Typhoon Sinlaku were used for verification. In order to approximate real conditions, scaling trees, electric towers and containers were set in the same places in the Erchung Floodway. Moreover, where the experimental water stage was lower than the real one, symmetric stones of 3 cm to 5 cm in size were set on appropriate locations in the riverbed or flood plain to increase roughness. The worst validations of Typhoon Jangmi occurred at cross-section K19A of the Keelung River and cross-section F01 of the Erchung Floodway. After tuning the riverbed roughness value, the Manning's n of the model, to fit the real water stage in the calibration process, the verification process was conducted. The results of model validation showed that the errors range from 0.0% to 4.0% for calibration and 0.0% to 5.9% for verification. All water stage errors were less than 4.0% for calibration and 5.9% for validation (Table 5-5).

Table 5.5. The physical model calibration and verification for water stages

Event		Typhoon Jangmi*			Typhoon Krosa**			Typhoon Sinlaku**		
Position		Exp.	Obs.	Error	Exp.	Obs.	Error	Exp.	Obs.	Error
Dahan Creek	T036A	5.33	5.33	0.0%	6.55	6.55	0.0%	5.07	5.02	1.0%
Hsindian Creek	S10A	5.93	5.93	0.0%	7.10	7.09	0.1%	5.49	5.45	0.6%
Erchung Floodway	T032	4.97	5.08	2.2%	6.00	6.09	1.5%	4.73	4.77	0.9%
	F009A	4.74	-	-	5.68	-	-	4.62	4.65	0.6%
	F008A	3.82	-	-	4.79	4.96	-3.5%	-	2.95	-
	F005A	3.00	-	-	3.98	3.80	4.6%	-	2.80	-
	F001	2.80	2.70	3.7%	3.58	3.70	3.4%	2.93	2.86	2.3%
Keelung River	K019A	3.78	3.63	4.0%	3.90	3.89	0.1%	4.33	4.09	5.9%
Tanshui River	T024A	3.57	3.54	0.8%	4.45	4.39	1.4%	3.54	3.45	2.6%
	T013	2.70	2.70	0.0%	3.42	3.42	0.0%	2.86	2.86	0.0%

Unit: meter.

* model calibration; ** model verification.

The results reveal that the physical model has realistic prediction capability that can be used to simulate different scenarios for testing the Taipei Flood Prevention System. Furthermore, according to the experimental results of the physical model, the Q_{200} of the Tanshui River increased from 14 300 m^3/sec (designed flood discharge) to 17,200 m^3/sec when the Q_{200} of the Erchung Floodway decreased from 9200 m^3/sec (designed flood discharge) to 6300 m^3/sec, but the water stage at T024A was 8.11 m, less than 8.40 m (designed flood stage).

Numerical Model Validation

Manning's n values, which are the major model parameter in the HEC-RAS model, needed to be verified. The process of verification contained three steps.

First, in order to determine the Manning's n values by comparing water stages between the physical model experiment results and the preliminary simulation results, the Tanshui River and the Erchung Floodway were separated and simulated as a single river under the given upper boundaries based on the result of the physical model experiment (the values were 17,200 m^3/sec and 6,300 m^3/sec for the Tanshui River and for the Erchung Floodway, respectively). Second, the Tanshui River and the Erchung Floodway were simulated as networking rivers to further adjust the preliminary Manning's n values by comparing the Q_{200}-Erchung Floodway (calibrated) and the Q_{100}-Erchung Floodway (verified), which were automatically estimated by HEC-RAS based on the energy conservation in accordance with the physical model experiment results (5400 m^3/sec for the Q_{100}-Erchung Floodway and 6300 m^3/sec for the Q_{200}-Erchung Floodway). Third, water stages were compared with the physical model experiment results again. If the error of the water stage was unacceptable, the Manning's n value of that cross section was adjusted, and then the process of verification returned to the previous step to check the value of bypass discharge until the bypass discharge and water stage errors were both acceptable.

The result shows that the bypass discharge errors were 0.2% for calibration and 2.0% for verification (Table 5-6). Under this prerequisite, all water stage errors were less than 3.7% for calibration and less than 3.4% for verification (Table 5-7). Therefore, the numerical model can be used to simulate the flow discharge of flood diversion and the water stage of each cross-section.

Table 5.6. The numerical model calibration and validation for bypass discharge

Return period (yr)	River	Flood flow (m^3/sec)		
		Physical model	Numerical model	error
200 (calibration)	Tanshui River (Taipei Br.)	17,200	17,086	0.2%
	Erchung Floodway	6,300	6,314	
100 (verification)	Tanshui River (Taipei Br.)	18,100	16,192	2.0%
	Erchung Floodway	5,400	5,508	

Table 5.7. The numerical model calibration and validation for water stages

Event Position		Q_{200} (calibration)			Q_{100} (verification)		
		Numerical model	Physical model	Error	Numerical model	Physical model	Error
Dahan Creek	T036A	9.68	9.62	0.6%	9.22	9.12	1.1%
Hsindian Creek	H010A	10.5	10.46	0.4%	10.19	10.32	-1.3%
Keelung River	K019A	7.58	7.58	0.0%	7.06	7.14	-1.1%
Erchung Floodway	T032	9.36	9.1	2.9%	8.92	8.68	2.8%
	F011	9.09	8.81	3.2%	8.66	8.39	3.2%
	F009A	8.78	9.01	-2.6%	8.35	8.59	-2.8%
	F008A	8.03	7.91	1.5%	7.55	7.49	0.8%
	F005A	7.66	7.58	1.1%	7.18	7.12	0.8%
	F001	7.24	6.98	3.7%	6.78	6.56	3.4%
Tanshui River	T024A	8.13	8.11	0.2%	7.69	7.64	0.7%
	T013	6.78	6.79	-0.1%	6.33	6.36	-0.5%
	T010	4.91	4.91	0.0%	4.6	4.61	-0.2%
	T000	2.3	2.3	0.0%	2.2	2.2	0.0%

Unit: meter.

Investigation of Currently Vulnerable Areas

According to the experimental results, the Q_{200}-Tanshui River increased from 14,300 m^3/sec (designed flood discharge) to 17,200 m^3/sec whereas the Q_{200}-Erchung Floodway decreased from 9,200 m^3/sec (designed flood discharge) to 6,300 m^3/sec, but the water stage at T024A was 8.11 m, and less than 8.40 m (designed flood stage).

This means that three areas would encounter flood inundation under Q_{200} and these are Shihtzutou (left bank from F001 to T013), Shezi (right bank from T013 to T019) and Wugu (left bank from F001 to F005A). They are currently vulnerable areas (Figure 5-10).

The levee of Shihtzutou, which is approximately 930 m long, is a gap in the Taipei Flood Prevention System. Moreover, the two areas of Shezi and Wugu were designed to serve as detention basins for containing flood volume in disaster mitigation. However, these two districts now have dense populations (population density was 20,154 persons/km^2 for Shezi and 2,270 persons/km^2 for Wugu in 2009 and have faced development pressure recently. The residents and the Taipei City Government cannot afford inundation of these areas.

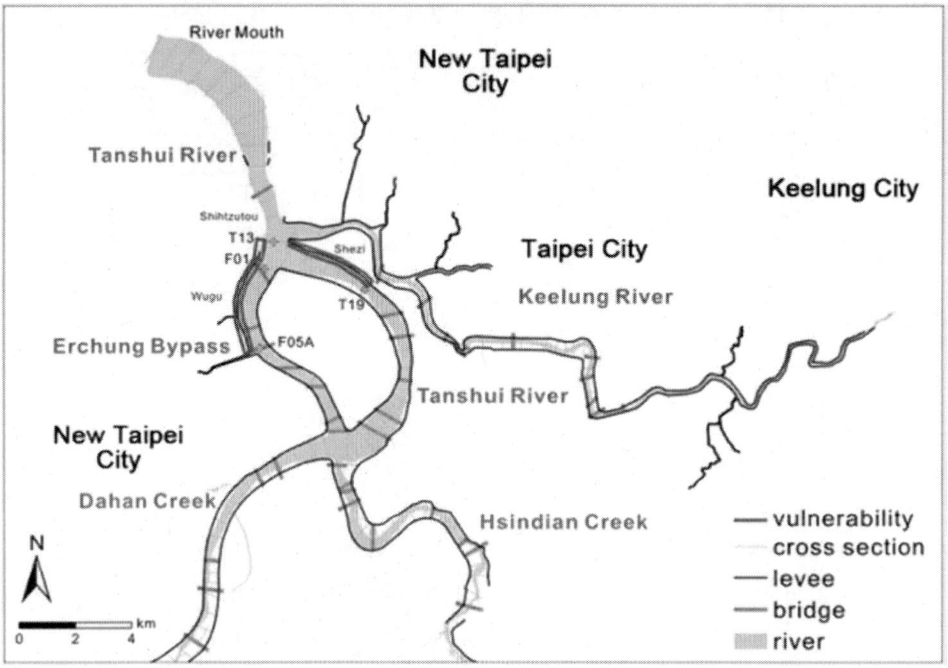

Figure 5.10.

DISCUSSION

Anthropogenic Effects

Riverbed Elevation

The riverbed alterations in the Erchung Floodway and in the Tanshui River both influence the strength of the Taipei Flood Prevention System. The bathymetry in our numerical model was further modified to measure the impact of riverbed alteration.

In 2010, the riverbed of the Erchung Floodway was higher than in 1986 (built up); this could be due to early industrial area development or environmental improvement of riparian landscape (the Taipei County Government, 2009). When the riverbed variations in the Erchung Floodway were restored from levels in 2010 to those in 1986, the Q_{200}-Erchung Floodway increased from 6,314 m^3/sec to 6,800 m^3/sec calculated by the numerical model. Meanwhile, the water stage at the Taipei Bridge declined from 8.13 to 8.09 m. An inverse relationship between the riverbed elevation within the Erchung Floodway and the Q_{200}-Erchung Floodway was found. In order to further investigate this impact, we divided the Erchung Floodway into three sections: the upper section (weir to F08A), middle section (F008A to F004) and lower section (F004 to F001). The simulated results showed that the Q_{200}-Erchung Floodway increased to 300 m^3/sec, 170 m^3/sec and 130 m^3/sec when the topographies of the upper, middle and lower sections were restored from 2010 to 1986 levels, respectively. Therefore, the influences of diverting the discharge and water surface elevation from riverbed elevation variation in the Erchung Floodway was: up-stream section (from the weir to Chung-Shang Bridge) > mid-stream section (from Chung-Shang Yuang-An Bridge to Bridge)> down-stream section (from Yuang-An Bridge to outlet gate).

On the other hand, the topography within the Tanshui River in the numerical model was modified as well. The simulated results showed that, when the topography was restored from 2010 to 1989 levels (the year of sand mining banned), the Q_{200}-Erchung Floodway decreased from 6,314 m^3/sec to 5,035 m^3/sec, but the water stage at T024A declined from 8.13m to 7.51 m. In addition, when the topography was restored from 2010 to 1970 levels (the year the bathymetry survey began), the Q_{200}-Erchung Floodway and the water stage at T024A were nearly the same (6,303 m^3/sec and 8.21 m). As mentioned above, the riverbed of the Tanshui River kept descending due to sand mining before 1989. Thus, water stages were lower and the Q_{200}-Erchung Floodway decreased. After banning sand mining, riverbed elevation stopped degrading and has continued rising until now. This caused the Q_{200}-Erchung Floodway to decrease from 6,350 m^3/sec in 1970 to 5,035 in 1989. The result indicates that the flood bypass capability recovers to 1970 before opening for sand mining activity.

Riverbed Roughness

The Manning's n parameter in the numerical model was modified to analyze the impact of riverbed roughness on the strength of the Taipei Flood Prevention System. The value of Manning's n of the Erchung Floodway was further modified to -50%, -25%, +25%, +50% and +100% of the original one. The simulated results showed that the riverbed roughness in the Erchung Floodway significantly affects the Q_{200}-Erchung Floodway. An inverse relationship between the riverbed roughness in the Erchung Floodway and the Q_{200}-Erchung Floodway was found (Figure 5-11). The Q_{200}-Erchung Floodway decreased from 6314 m^3/sec to 8335m^3/sec when the riverbed roughness in the Erchung Floodway increased 100%. This means that it would be more difficult for floods to enter the Erchung Floodway and consequently flow toward the Tanshui River under high riverbed roughness of the Erchung Floodway. The water stage at T024A would exceed the designed flood stage if the riverbed roughness in the Erchung Floodway increased by over 25%. This would not be conducive to flood prevention. Conversely, if the riverbed roughness in the Erchung Floodway decreases, more floods could pass by the Erchung Floodway. The Q_{200}-Erchung Floodway would increase by 1100 m^3/sec and the water stage at T024A would decline by 0.09 m if the riverbed roughness of the Erchung Floodway decreased by 50%. Moreover, the Erchung Floodway was further divided into the upper (the weir to F008A), middle (F008A to F004) and lower (F004 to F001) sections to reduce the riverbed roughness. The simulated results showed that the impacts of riverbed roughness in the Erchung Floodway are as follows: up-stream section (from the weir to Chung-Shang Bridge) > mid-stream section (from Chung-Shang Yuang-An Bridge to Bridge) > down-stream section (from Yuang-An Bridge to outlet gate) (Figure 5-12).

On the other hand, the values of Manning's n of the Tanshui River in the numerical model were modified to -50%, -25%, +25%, +50% and +100% as well. The simulated results showed that the riverbed roughness in the Tanshui River significantly affects not only the Q_{200}-Erchung Floodway, but also the water stage at T024A (Figure 5-13). This means that the riverbed roughness of the Tanshui River is a sensitive environmental factor influencing the Taipei Flood Prevention System. The Q_{200}-Erchung Floodway increased to about 800 m^3/sec, 1600 m^3/sec and 3000 m^3/sec when the riverbed roughness increased 25%, 50% and 100%, respectively. However, the water stage at T024A would exceed the designed flood stage if the riverbed roughness increased by over 25%.

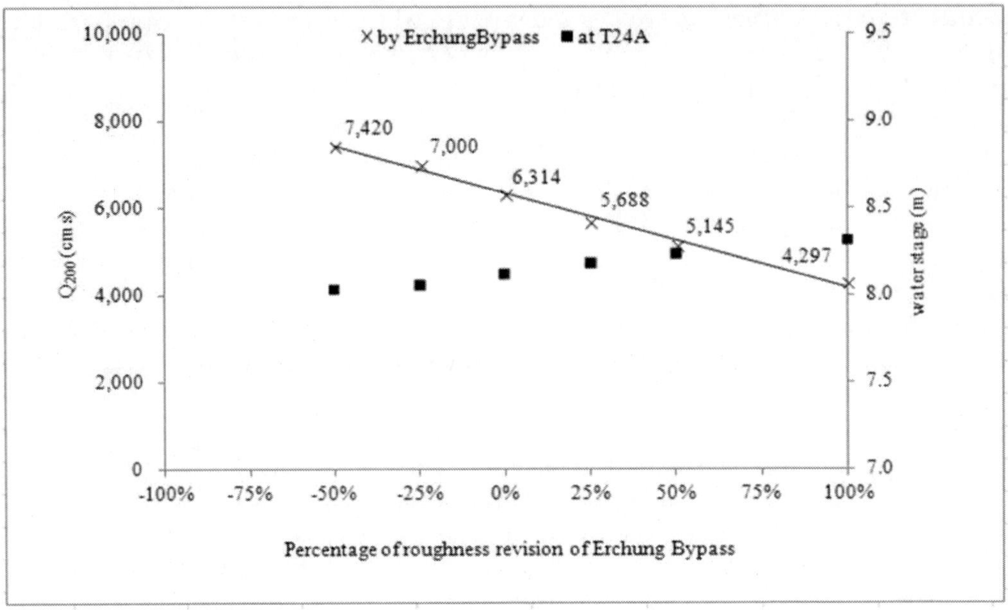

Figure 5.11.

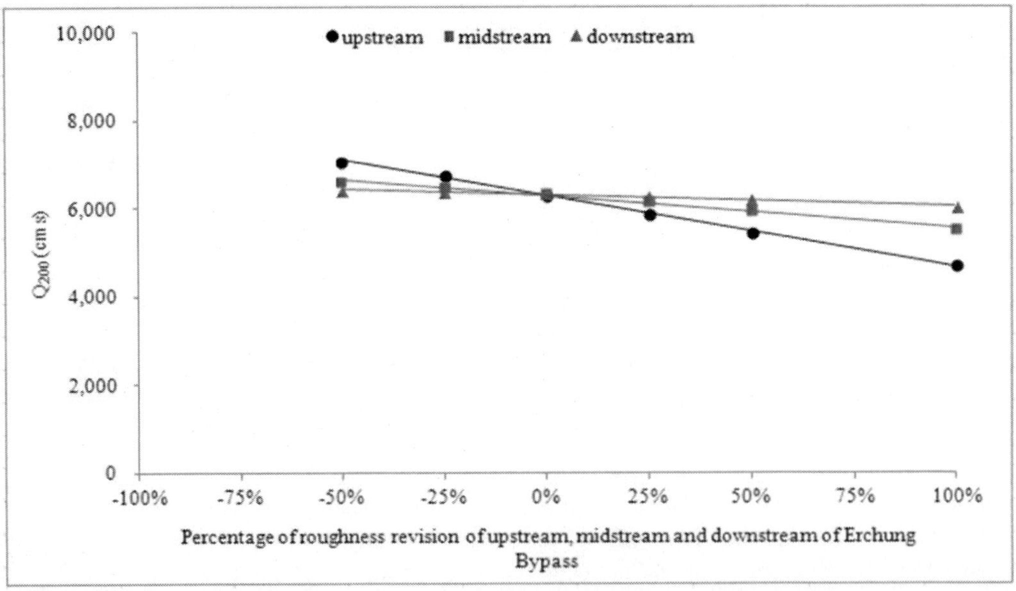

Figure 5.12.

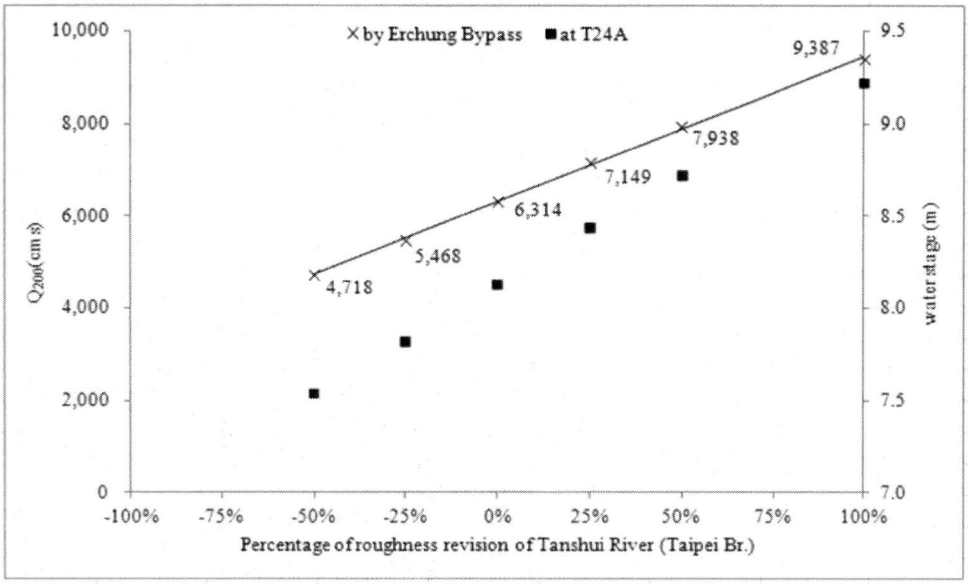

Figure 5.13.

Climate Change Threats

Effect of Sea Level Rise

Five scenarios were set to analyze the impact of water stage at the river mouth (downstream condition of the Taipei Flood Prevention System): 2.30 m, 2.90 m, 3.50 m, 4.03 m and 5.03 m at T000.

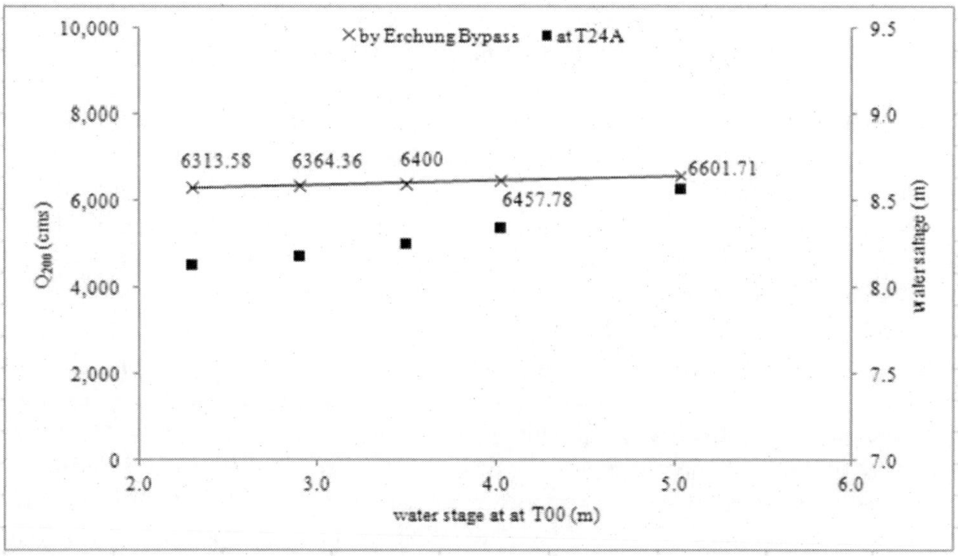

Figure 5.14.

As shown in Figure 5-14, a rising water stage at T000 (river mouth) would not only increase the Q_{200}-Erchung Floodway, but also raise the water stage at T024A. A high water stage at T024A (the bottleneck) would increase flood protection pressure on the Taipei Flood Prevention System, although the Q_{200}-Erchung Floodway may decrease. For instance, the water stage at T024A was 8.57 m when the water stage was set as 5.03 m and higher than the 8.40 m (the designed flood stage).

Increase of the 200-Year Return Period Flood Discharge

Considering extreme rainfall due to climate change, the 200-year return period flood could increase from 25,000 m^3/sec to 28,300 m^3/sec in the Taipei Flood Prevention System (Water Resources Planning Institute, 2010). The Q_{200}-Erchung Floodway increased by about 1200 m^3/sec (from 6,456 m^3/sec to 7,659 m^3/sec) when increasing the 200-year flood return period. However, the water stage at T024A rose to 9.00 m, which was higher than that of the designed flood stage. The Taipei Flood Prevention System would encounter a severe challenge in this scenario.

CONCLUSION

The Tanshui River basin has frequent flooding and inundation along riversides and surrounding lowlands due to heavy precipitation associated with storms or typhoons. It took 37 years (from 1963) to complete the Taipei Flood Prevention System. Due to rapid urbanization, land development and natural alteration by human activities have occurred. The riverbed elevation and roughness changes threatened the strength of the Taipei Flood Prevention System. In this chapter, the current protection criteria from anthropogenic effects or climate change threats were examined and analyzed by numerical or physical modelling to provide useful information for river management.

Based on the results of the sensitivity analysis, six factors would raise the water stage in the Tanshui River: (1) riverbed accretion in the Erchung Floodway; (2) riverbed accretion in the Tanshui River; (3) increasing riverbed roughness in the Erchung Floodway; (4) increasing riverbed roughness in the Tanshui River; (5) increasing the 200-year flood return period; (6) increasing the water surface elevation of the river mouth. Indeed, increasing riverbed roughness in the Erchung Floodway by over 50%, increasing riverbed roughness in the Tanshui River by over 25%, and increasing the 200-year return flood discharge by over 13% would raise the water stage at T024A to exceed that of the designed flood stage. For example, the simulation resulting from climate change effect indicates that the increase of the 200-year return flood discharge causes three vulnerable locations in the Taipei Flood Prevention System mentioned previously. Increasing riverbed roughness in the Erchung Floodway or in the Tanshui River could have resulted from anthropogenic change or natural succession, such as riverside park construction and planting vegetation. This can be improved through appropriate river management strategies.

Two cost-effective strategies to restore the Q_{200}-Erchung Floodway and decrease water stages were proposed according to the simulation results: a) maintaining the riverbed elevation of the Tanshui River, and b) decreasing the riverbed roughness in the Erchung Floodway. In the first strategy, the measured riverbed elevations in the Tanshui River in 2010 were getting close to those in 1970 and continued to rise due to the banning of sand mining. Besides, the riverbed elevations would rise by about 0.3 m and 3.05 m in 12 to 48 years of sand-bypass operation in the Shihmen Reservoir (Water Resources Agency MOEA, 2011). Therefore, the riverbed elevation of the Tanshui River should be monitored and silt dredging should be carried out to reduce riverbed elevation if it becomes necessary. In the second strategy, the riverbed roughness of sections in the Erchung Floodway should be managed. Most of the lands in the upper and middle sections are covered by farmland, parking lots and rinks (the Taipei County Government, 2010). Because the upper and middle sections significantly affect the Q_{200}-Erchung Floodway, we suggest that the riverbed roughness should be decreased by at least 25%. It should be noted that river velocity could increase and might damage infrastructure, such as bridges and levees, after decreasing riverbed roughness. In the lower section, most of the land is covered by grass, wetland or water. The riverbed roughness has been low. For instance, the value of Manning's coefficient varies between 0.025 and 0.035 where the land is covered by grass (Chow, 1973). We suggest that the lower section be maintained as it is currently because the effect of improvement would be limited. In addition, there are a number of constructed wetlands in operation and *Mortonagrion hirosei*, an endangered damselfly species, inhabits the lower section (the International Union for Conservation of Nature, IUCN).

On the other hand, the mangrove *Kandelia obovata* is spreading rapidly in the Tanshui River and has a competitive growth advantage within approximately 14.5 km (T21) of the river mouth along the Tanshui River. The roughness would obviously increase due to mangrove tree invasion into the mudflats. Mangrove invasion is expected to increase due to salt intrusion enhanced by rising sea levels, and riverbed roughness within the Tanshui River would consequently increase (Shih et al. 2011). As Yang et al. (2013) mentioned: "These mangrove forests do not only need protection projects but also require management projects." Therefore, the riverbed elevation and roughness within the Tanshui River should be monitored. To reduce flood risk, silt dredging or mangrove trimming should be carried out regularly to reduce riverbed elevation and roughness, respectively.

The physical model was further tested to investigate the function of the Taipei Flood Prevention System by reducing riverbed roughness through removing half of the original number of trees from the weir to F08A. Furthermore, the riverbed elevation near the weir was dredged by two to four meters in 2011and the mudflat elevation of the Huajiang Wetland was rehabilitated from around 3.0 m to 0.5 m for common teals (*Anas crecca*). These two latest alterations were included in the physical model modification. The results showed that the water stage at T024A declined by 0.1 m (from 8.11 to 8.01 m), whereas the Q_{200}-Erchung Floodway increased by 275 m^3/sec. In other words, reducing riverbed roughness in the upper and middle Erchung Floodway sections is a workable and cost-effective management strategy to restore and maintain the function of the Taipei Flood Prevention System without raising the height of levees.

We conclude that the degrading process of the Taipei Flood Prevention System due to rapid urbanization including river sand mining and the cost-effective mitigation strategies including river roughness and riverbed elevation control are meaningful lessons especially for developing countries. However, water stage variations at the river mouth and the increased discharge of the 200-year return flood period may be induced by rising sea level and climate change, respectively. These two phenomena are not easy to predict and overcome. It means that a system in which levees and pumps are installed to prevent urban flooding, such as the

Taipei Flood Prevention System, could encounter challenges in the near future. It is suggested that a comprehensive and effective evacuation system be established and launched as soon as possible.

ACKNOWLEDGMENT

We appreciate the Tenth River Management Office, Water Resource Agency for funding support and providing the data of riverbed bathymetry and relevant hydrological data. This chapter benefited from the discussion with Dr. Sheng-Chi Yang and Dr. Huei-Tau Ouyang.

REFERENCES

Botzen, W. J. W., van den Bergh, J. C. J. M., Bouwer, L. M., (2010). Climate change and increased risk for the insurance sector: a global perspective and an assessment for Netherlands. *Nat. Hazards*, 52:557-598.

Chen, A. S., Hsu, M. H., Tsng, W. H., Huang, C. J., Yeh, S. H., Lien, W. Y., (2006). Establishing the database of inundation potential in Taiwan. *Nat. Hazards*, 37, 107–132.

Chow V. T., (1973). Open-Channel Hydraulics, McGraw Hill International Editions, Singapore.

Dang N. M., Babel M. S., Luong H. T., (2011). Evaluation of food risk parameters in the Day River Flood Diversion Area, Red River Delta, Vietnam, *Nat. Hazards*, 56,169–194.

Hsieh, L. S., Hsu, M. H., Li, M. H., (2006). An assessment of structural measures for flood-prone lowlands with high population density along the Keelung River in Taiwan. *Nat. Hazards,* 37, 133–152.

Huang, S. L., Hsu, W. L., (2003). Materials Flow Analysis and Emergy Evaluation of Taipei's Urban Construction. *Landscape and Urban Planning*, 63, 61-74.

IPCC, (2001). Climate Change 2001: Impacts, adaptation and vulnerability—Contribution of Working Group II to the Third Assessment Report of the Intergovernment Panel on Climate Change. Cambridge University Press, Cambridge, UK.

Kundzewicz, Z. W., Ulbrich, U., Brucher, T., Graczyk, D., Kruger, A., Leckebusch, G. C., Menzel, L., Pinskwar, I., Radziejewski, M., Szwed, M., (2005). Summer Floods in Central Europe – Climate Change Track? *Natural Hazards*, 36, 165-189.

Lehner, B., Doll, P., Alcamo, J., Henrichs, T., Kaspar, F., (2006). Estimating the impact of global change on flood and drought risks in Europe: A continental, interated analysis. *Climate Change*, 75, 273-299.

Lowe, A. S., (2003). The federal emergency management agency's multi-hazard flood map modernization and the national map. *Photogrammetric Engineering and Remote Sensing,* 69(10),1133–1135.

Ministry of Economic Affairs, (1973). Proposal for Taipei Flood Prevention, Ministry of Economic Affairs, Taiwan (in Chinese).

Murphy, D., (2003). Strategy for flood risk management (2003–2008). Environment Agency, UK.

Pan, T. Y., Chang, L. Y., Lai, J. S., Chang, H. K., Lee, C. S., Tan, Y. C., (2012). Coupling typhoon rainfall forecasting with overland-flow modeling for early warning of inundation, *Natural Hazards*, DOI 10.1007/s11069-011-0061-9.

Shih, S. S., Yang, S. C., Lee, H. Y., Hwang, G. W., Hsu, Y. M., (2011). Development of a Salinity-Secondary Flow-Approach Model to Predict Mangrove Spreading. *Ecological Engineering,* 37, 1174-1183.

Taipei County Government, (2009). Master plan for establishing Taipei Metropolitan Park in the Erchung Floodway, Taipei County Government, Taiwan (in Chinese).

Taipei County Government, (2010). The evaluation and assessment of flood mitigation effects after Wugu dyke construction, Taipei County Government, Taiwan (in Chinese).

U.S. Army Corps of Engineers, (2010a). HEC-RAS Hydraulic Applications Guide: Version 4.1. Insititute for Water Resource, Hydrologic Engineering Center, Davis, CA.

U.S. Army Corps of Engineers, (2010b). HEC-RAS Hydraulic User's Manual: Version 4.1. Insititute for Water Resource, Hydrologic Engineering Center, Davis, CA.

Water Conservancy Bureau, (1972). Report on the work of the technology group of Taipei Flood Prevention 03-(Flood)-11, Water Conservancy Bureau, Taiwan (in Chinese).

Water Resources Agency MOEA, (2011). Impact evaluation of sediment released from Shih-men reservoir on downstream river morphology, Water Resources Agency MOEA, Taiwan (in Chinese).

Water Resources Bureau MOEA,(1996). Experimental report for hydraulic model of the Tanshui River establishment and Taipei Flood Prevention performance. Water Resources Bureau MOEA, Taiwan (in Chinese).

Water Resources Planning Commission MOEA, (1966). Report of the Tanshui River hydraulic model experiments, Water Resources Planning Commission MOEA, Taiwan (in Chinese).

Water Resources Planning Commission MOEA, (1970). Review for Taipei Flood Prevention System appendix 7: meeting minutes, Water Resources Planning Commission MOEA, Taiwan (in Chinese).

Water Resources Planning Commission MOEA, (1974). Experimental report for arrangement and stability of the Tanshui River Bypass entry, Water Resources Planning Commission MOEA, Taiwan (in Chinese).

Water Resources Planning Commission MOEA, (1980). Report on the hydraulic model experiments of the Erchung Floodway entry, Water Resources Planning Commission MOEA, Taiwan (in Chinese).

Water Resources Planning Commission MOEA, (1983). Related projects to preliminary Taipei Flood Prevention System: Presentation on the hydraulic model experiments of the Erchung Floodway, Water Resources Planning Commission MOEA, Taiwan (in Chinese).

Water Resources Planning Institute WRA, (2010). Hydrological review of Tanshui River system, Water Resources Planning Institute WRA, Taiwan (in Chinese).

Yang, S. C., Shih, S. S., Hwang, G. W., Adams, J. B., Lee, H. Y., Chen, C. P., (2013). The Salinity Gradient Influences on the Inundation Tolerance Thresholds of Mangrove Forests. *Ecological Engineering*, 51, 59-65.

In: Flooding
Editors: M. R. Motsholapheko and D. L. Kgathi
ISBN: 978-1-63117-607-4

Chapter 6

THE INFLUENCE OF RISK PERCEPTION AND ATTITUDE ON THE DECISIONS TO ADOPT RESIDENTIAL FLOOD INSURANCE: EVIDENCE FROM QUEENSLAND, AUSTRALIA

*Alex Y. Lo**
Griffith School of Environment, Griffith University, Australia

ABSTRACT

In 2011 in Queensland Australia, floods created considerable financial pressure on regional governments. Many affected households suffered severe economic losses as they did not have flood cover on their home insurance policies. The absence of flood insurance could pose threats to fiscal health and has risen to the national policy agenda. This study contributes to the debates by identifying key subjective factors associated with non-insurance. It is based on a social survey involving a total of 501 residents of Brisbane, the Gold Coast, and the Sunshine Coast. A significant minority of respondents (43.8%) reported to have no flood cover. Perceived flood risk is not statistically related to the likelihood of non-insurance. This means that non-insured households are not restricted to those who are unaware of the flood risks confronting them. The insured individuals are better educated and tend to recognize the role of flood insurance in financially protecting the household. Non-insurance is also associated with the expressed preference for government compensation over insurance. The study offers two main insights for policy-makers, floodplain managers and insurers. First, raising risk awareness is unlikely to be sufficient to improving the uptake of flood insurance. Second, managing the public expectations about disaster relief may have positive impacts.

Keywords: flood insurance; non-insurance; flooding; risk perception; public attitude; Queensland

* Email: alex.lo@griffith.edu.au.

INTRODUCTION

Rare but large-scale weather extremes could severely disrupt public finance. Governments of affected areas are confronted with enormous costs of rescue operations, loss of human life, business disruption, community rebuilding or relocation (The World Bank, 2010). Unanticipated natural perils may even trigger compensation claims against faulty land and flood management decisions made decades ago (McDonald, 2010). Rising frequency and severity of extreme weather events threaten to increase the number of large payouts from the public purse. Governments are at pains to spare such unusually large amounts of contingency funds to prepare for rare catastrophes (Handmer, 2002; McDonald, 2010). In response many government agencies have included natural disaster insurance schemes in their risk management plans (The World Bank, 2010). The most prominent example is the U.S.'s National Flood Insurance Program, which has been operating since 1968.

Australia does not yet have a national approach for regulating the provision of commercial flood insurance in the country, despite many of its urban and regional settlements being subject to severe flooding (Handmer, 2002; 2008; Handmer and Smith, 1989; Smith and Handmer, 2002). The problem of the lack of national coordination exploded in the aftermath of the recent severe flooding events in Queensland. Extensive floods hit Queensland in late December 2010 and early January 2011. Prolonged and extensive rainfall over large areas of Queensland, coupled with already saturated catchments, led to flooding of historic proportions in the State (Queensland Government, 2012). As a consequence, more than 136,000 homes and businesses were affected and 33 people lost their lives (Queensland Government, 2011, p. 16; 2012, p. 32). The floods severely damaged the regional economy at an unprecedented scale (Lim, et al., 2012; Thomas, et al., 2011). It was the second largest event in terms of catastrophe insurance claims since 1967 (Reserve Bank of Australia, 2011, p. 39). The resulting economic losses amounted to AUD$5.8 billion (Queensland Government, 2011, p. 3).Consequently the government was forced to seek relief from controversial ad hoc measures, such as the Temporary Flood and Cyclone Reconstruction Levy introduced after the 2011 Queensland floods.

In anticipation of a changing climate, governments at various levels pledge to enhance community resilience by means of natural disaster insurance, among other measures (McDonald, 2010; Phelan, et al., 2011). In Australia, a series of federal reviews and inquiries were conducted to examine current institutional arrangements for flood insurance and the practice of the insurance industry. In the months after the 2010/11 floods, the Australian government called for a review of issues about natural disaster insurance, with a focus on issues about flood coverage. It is supported by the Treasury and known as the 'Natural Disaster Insurance Review' (NDIR) (Australian Treasury, 2011). There were also other reviews and public inquires established at the federal level by the Australian Treasury and the House of Representatives. Later in 2011, Australia's Productivity Commission invited individuals and organizations to respond to a public inquiry into the barriers to climate change adaptation in Australia (Productivity Commission, 2013). This public inquiry has extensive coverage on the provision of natural disaster insurance.

All of these reviews and public inquiries have addressed the enduring problem of non-insurance[*] in Australia. In the absence of adequate federal sponsorship, only a very small proportion of those households at risk of flooding took out flood cover (Handmer and Smith, 1989). By 2006, only 3% of the policies purchased in Australia covered flood, although it has risen to 54% in 2010 (Insurance Council of Australia, 2011, p. 8). Despite the insurance coverage rate has gone up nationally, geographic variations remain significant and warrant attention.

The situation in Queensland is particularly alarming because many Queenslanders are not insured for flooding (Tooth, 2012). In Australia, flood cover is typically excluded from home and contents policies, or where flood cover is offered, in many cases the policyholder can and *does* opt-out (Australian Government, 2011). Some of the Queensland policyholders were surprised and disappointed when they came to realize that their insurance policies did not cover the type of inundation that created their losses in the 2011 devastating floods (Australian Government, 2011; Carter, 2012; Mortimer et al., 2011). The tendency to opt-out and misunderstanding of policy coverage fly in the face of the knowledge that a large number of human settlements in Queensland are located at flood-prone areas. Improving the uptake of flood insurance in the State is undoubtedly an imperative.

The government and the industry have suggested various reasons for the lack of motivation to insure against flood risks, including low awareness of risk, moral hazard, affordability, institutional trust, etc. However, the ways in which these factors influence household decision are not well understood. As pointed out by the Productivity Commission (2011, p. 231), there have been few Australian studies of non-insurance. Nonetheless, Tooth's (2012) recent research could offer some useful insights. For instance, one-third of households residing in high risk areas (including Queensland) did not have flood cover (Tooth, 2012). The proportion of those without flood insurance is higher in Queensland than any other State or Territory in Australia (Tooth, 2012, p. 29) – even nearly one year after the 2011 Queensland floods[†] which presumably had led to increased awareness of flood risk among Queenslanders. These observations allow a conjecture that seemingly important factors, such as risk awareness, may have little effect on the insurance purchase decision. An examination of those possible factors is needed.

This study aims to identify key subjective factors associated with non-insurance. Risk perception and community attitude toward institutional arrangements were solicited and systematically analyzed in terms of their statistical relationships with the likelihood of non-insurance. Findings could provide insights into the household decision of purchasing flood insurance. They could then help devise strategies for promoting voluntary adoption of flood insurance, which ultimately could help ameliorate financial pressures on governments and communities in times of major flooding events. The study is of practical interest for policy-makers, floodplain managers and the insurance industry who are concerned about the low penetration of flood insurance in Australia, notably Queensland.

The study is based on a social survey administered in the major cities of South East Queensland: Brisbane, the Gold Coast and the Sunshine Coast, which are among the most vulnerable areas in Australia, being exposed to escalating risks of flooding (Australian

[*] Non-insurance is defined as a situation 'where insurance is not taken out, or does not cover a relevant hazard such as flooding' (Productivity Commission, 2011, p. 231). This definition is adopted in this paper with specific reference to flooding.

[†] The household survey conducted by Tooth (2012, p. 15) was conducted in mid-December 2011.

Government, 2009) and in need of a more robust system of insurance to shield their inhabitants from future natural hazards. This paper reports results of the survey. The next section introduces the study areas, survey instrument and sampling method, followed by the results of the survey. Practical implications are discussed in the conclusions.

METHODS

Study Areas

This research was conducted in the south-eastern cities of Queensland affected by the 2011 floods to different extents. Survey sites include the Gold Coast Council Region, the Sunshine Coast Council Region, and the Greater Brisbane Region (i.e., Brisbane, Ipswich, Logan, Redland and Moreton Bay Council areas). According to Australian Bureau of Statistics (ABS) 2011 Census data, these regions altogether account for 64.3% of the total population of Queensland, or 13.0% of national population.

In January 2011 Brisbane experienced its second highest flood since the beginning of the 20th century. Major flooding occurred in the Lockyer, Bremer and Brisbane Rivers. The Bremer River level at Ipwich peaked at 19.25 metres and the Brisbane River level at Brisbane peaked at 4.46 metres (Dowling, 2012). The Gold Coast is ranked as the most flood prone local council in Australia(Smith, 2002). Recent studies confirm that the Gold Coast and the Sunshine Coast remain among the Queensland's local government areas at the greatest risk of inundation, exacerbated by their rapid population growth and intensive development along the shore over the last few years (Department of Climate Change, 2009; Roche, et al., 2010). Extreme weather events are likely to occur more frequently. Although climate change is expected to bring lower rainfall to South East Queensland, projections suggest an increase in extreme rainfall events for large areas of the region (Queensland Government, 2011). The study sites are characterised by increasing flood risks coupled with continued population growth.

Questionnaire Design

A structured questionnaire was designed for this research to collect primary data from South East Queensland households. The survey questions used can be grouped into four categories: insurance status, risk perception, attitude toward flood insurance and socio-economic characteristics of respondents.

In the first part of the questionnaire, respondents were asked if they currently had home and/or contents insurance on the residential property they were living. Those who returned an affirmative response were then probed whether or not they had flood cover on that insurance.

Then, respondents rated their flood risks using a five-point scale with options ranging from 'No risk' to 'Extreme risk'. The survey question was 'How would you assess the risk of flooding that the home or property you are currently living in is exposed to?' It was included to address practical considerations. Official reports typically make the assumption that if individuals had known of their risk exposure, they may have taken up insurance coverage

(House of Representatives Standing Committee on Social Policy and Legal Affairs, 2012, p. 37-50, cited in Carter, 2012). The Productivity Commission (2011) has also drawn a link between the individual's response to increased flood risk and their estimation of the probability of flooding. However, empirical support for this assumption is not unequivocal. Although Kunreuther (1996) has presented affirmative findings, other researchers such as Baumann and Sims (1978) and Hung (2009) fail to offer clear evidence. Further evidence is needed to justify the standard assumption.

Attitudes towards various aspects of flood insurance were gauged by six questions. These questions generally pertain to the extent to which the respondent affirms the role of flood insurance and the ways in which it is provided. First, respondents assessed the importance of flood insurance in providing financial security to the community and reflected on their responsibility for insuring. These survey items were included because of their clear intuitive relevance to the insurance decision. Also, attitude toward government providing *ex post* compensation to flood victims was also measured. The Productivity Commission and the NDIR Panel have expressed concerns about the government's promises to offer generous compensation to flood victims. One of the justifications is that knowledge of guaranteed disaster relief may create perverse incentives to private efforts on risk mitigation (Productivity Commission, 2011, p. 238) -a form of 'moral hazard' (Kriesel and Landry, 2004; McDonald, 2010; Mortimer, et al., 2011). This item, labelled as 'Compensation', was included to assess its possible effect.

Another item, labelled as 'Insurer', assessed trust in insurers. The issue of trust has been raised and described by the Productivity Commission (2011, p231) in terms of 'negative past experiences with insurance claims'. Tooth (2012) had included perceived reliability of insurers in his survey but did not examine whether or not it would contribute to a skeptical view toward insuring against flood risks. The present study attempted to explore this possibility. Moreover, both the Productivity Commission and the NDIR panel have indicated a preference for strengthening regulation of the insurance market and the use of price signals to help the household to withstand the impacts of flooding. The remaining two survey questions were constructed to measure the public attitudes toward the role of market and price signals in an attempt to assess the effects of these beliefs on the tendency for insurance.

Socio-economic information of the respondents and their households were collected both at the beginning and the end of the questionnaire. These included household size, gender, age, education, and household income. Reported household size was used for initial screening of respondents. Sampling procedures are introduced and explained in the following section.

Sampling Method

The main survey was conducted during 16 May and 20 May 2012, preceded by pilot testing on 14 May 2012 to allow refinements to the survey instrument. A professional market research company was appointed to undertake the pilot and main surveys using the Computer Assisted Telephone Interviewing (CATI) technique. Sample was selected using a Random Digit Dialling (RDD) approach for households with landlines, stratified by telephone exchange prefix. It was supplemented with 10%-15% of residential mobile phone numbers selected from the Electronic White Pages (EWP) so that mobile numbers could be stratified by region (as these numbers have a residential address attached to them).

Quotas were used in the selection of household for the telephone interview. Since the survey sought to contact the person in the household who was the most knowledgeable about their household insurance, the introductory text asked to speak to the person in the household mainly responsible for payment of the major household bills. In order to obtain a representative sample of households, sample quotas were set by the number of people living in each household per region, based on the latest available census data (2006). The number of people in household was obtained from each contacted household who agreed to participate, and used for initial screening in accordance with the set quotas.

RESULTS

A total of 501 household representatives completed the questionnaire, at the response rate of 20.2%, of which 301 resided in the Greater Brisbane Region, 100 in the Gold Coast Council region, and 100 in the Sunshine Coast, yielding the response rates of 21.0%, 15.5%, and 24.9%, respectively.

Household Characteristics

Table 1 displays the percentage of questionnaires completed for each household category and the corresponding census statistics. It shows that the sample consisted of fewer households with only one person (18%) than actually resident in the regions (22.5%) – the Greater Brisbane accounted for the discrepancy. About 38% of the households surveyed across the three regions had two members only, slightly above the regional estimate of 35.6%. Families with four members or more were over-represented in the survey by a very small margin, i.e., 26.8%, comparing to the regional estimate of 25.4%. The quota sampling method managed to produce a representative sample in terms of household composition.

Table 1. Percentage of responses by household composition against census data

	Region		Number of persons per household				
			1	2	3	4	5 or more
Household (%)	Greater Brisbane	Survey	10.4	20.4	10.4	11.8	7.2
		Census	15.4	23.7	11.5	11.3	7.2
	Gold Coast	Survey	4.8	8.2	3.2	2.4	1.4
		Census	4.5	7.3	3.2	2.8	1.6
	Sunshine Coast	Survey	2.8	9.4	3.8	2.6	1.4
		Census	2.6	4.6	1.7	1.6	0.9
	Total	Survey	18.0	38.0	17.4	16.8	10.0
		Census	22.5	35.6	16.4	15.7	9.7

Source: Australian Bureau of Statistics 2006 Census.

Descriptive statistics of other socio-economic variables are listed in Table 2, including household income, gender, age, and education. About 31.1% of respondents had an annual household income below AUD$50,000, 47.3% earned AUD$50,000 to AUD$150,000 a year, and 12.2% had more than AUD$150,000. Less than half (44%) of respondents were male.

More than 70% of respondents were aged above 45. Nearly 41% held a university degree or higher. Mature and educated individuals were over-represented in the survey, possibly because the sampling strategy selected household heads for the survey.

Table 2. Percentage of responses by gender, age, education, and household income

Socio-economic characteristics	Category	Respondent (%)
Gender	Male	44.3
	Female	55.7
Age	18 - 30	3.2
	31 - 45	23.2
	46 – 60	35.9
	Over 60	37.7
Highest level of education completed	No formal qualifications	7.8
	Year 10 or equivalent	12.0
	Year 12 or equivalent	18.0
	Trade / apprenticeship	4.2
	Certificate / diploma	17.2
	University degree	28.7
	Higher University degree	12.2
Total Annual Household Income	Under AUD50,000	31.1
	AUD50,000 - 100,000	27.7
	AUD100,001 - 150,000	19.6
	AUD150,001 or more	12.2
	Refused to answer	9.4

The vast majority of respondents (83%) had both home and contents insurance policies, slightly lower than what was founded nationally (88%) by Tooth (2012). In this survey, 43.8% of respondents claimed to have no flood cover on their current policies or not holding a home insurance policy at all. The percentage of policyholders in each study area is shown in Table 3.

In Brisbane and the Gold Coast respectively, 43.8% and 41.9% of the residents surveyed had no flood cover. A higher proportion was recorded in the Sunshine Coast – nearly 46.2% were not protected by flood insurance. Note that the 81 respondents who did not indicate whether their policies have flood cover or not were excluded from Table 3 and from further analysis.

Table 3. Percentage of responses reported to have no flood insurance by study area

Insurance status	Greater Brisbane Region	Gold Coast	Sunshine Coast	Total
No flood cover	43.8	41.9	46.2	43.8
Have flood cover	56.2	58.1	53.8	56.2
Total	100	100	100	100

Note: 'No flood cover' includes absence of flood cover on current policies as well as absence of home insurance. 'Don't know / cannot recall' were excluded.

Risk Perception

Not many households reported living in an area of high flood risk. Nearly 60% of them felt that they were not exposed to any risk of flooding (Table 4). About one-third (32.1%) believed that they faced low flood risk. Less than 10% reported exposure to medium, high, or extreme risks. There are some discrepancies across the three study sites. More than half of the respondents in the Gold Coast and Sunshine Coast assessed their flood exposure as being low to extreme, whereas only one-third of those in the Greater Brisbane Region made the same assessment.

Table 4. Percentage of responses by perceived flood risk

Perceived flood risk of current home/property	Greater Brisbane Region	Gold Coast	Sunshine Coast	Total
No Risk	65.9	48.0	46.5	58.4
Low Risk	27.8	39.0	38.4	32.1
Medium Risk	4.0	10.0	12.1	6.8
High Risk	1.0	2.0	3.0	1.6
Extreme Risk	1.3	1.0	0.0	1.0
Total	100	100	100	100

Note: Medium, High and Extreme risk categories were merged into one group in subsequent analysis because of low frequencies.

Attitude toward Flood Insurance

In general, respondents held positive views towards flood insurance. As shown in Table 5, 70.7% agreed[‡] that flood insurance is important because it could provide financial security for their household. Three quarters of respondents (75.9%) accepted the responsibility for insuring their home against the risk of natural disaster. The level of distrust in insurers was fairly high. Less than half (42.6%) expressed concerns about the failures of insurance companies to recover the insured losses of flood victims, an issue that led to public backlash across Queensland and the whole country in the aftermath of the 2011 flooding (Connolly, 2011).

Very few respondents (13%) considered flood insurance as not necessary because disaster relief was seen as a substitute. The majority recognized the necessity of this risk-sharing device. More than half (58.2%) of the respondents affirmed the role of free market in offering affordable insurance products. Just over 50% believed that premiums could indicate the level of flood risk accurately. The strength of these beliefs was modest.

[‡] All percentages reported in Section 3.3 refer to the proportion of respondents who indicated 'Somewhat agree' or 'Strongly agree' to the relevant survey statements.

Table 5. Percentage of responses to statements about attitude towards flood insurance

Item	Survey statement	Percentage of 'Strongly agree' or 'Somewhat agree' responses	Mean#
Security	Flood insurance is important as it provides financial security to my household	70.7	3.90
Compensation	Flood insurance not necessary as government should provide compensation to victims of flooding	13.0	1.91
Responsibility	Getting insurance against the risk of natural disaster is my responsibility	75.9	4.13
Insurer	Flood insurance is useless as insurance companies have all sorts of excuses to decline requests for compensation for my losses	42.6	3.14
Market	Free markets are important as they can operate to give us lower premiums for flood insurance	58.2	3.71
Price Signal	Changes in insurance premiums give signals in the market place about the level of flood risk to which people are exposed	50.5	3.39

#Measured on a five-point Likert scale (min = 1, max = 5)

Predicting Non-Insurance

The data were further analyzed using logistic regression. The stepwise regression analyses included non-insurance as a dependent variable and a suite of household characteristics and subjective factors, all reported in the preceding sections, as explanatory variables. Only three of them indicated statistical significance, including Education, Security, and Compensation, as reported in Table 6. Insignificant variables were excluded from the final model.

Table 6. Binary logistic model for non-insurance

Variable	B	S.E.	Wald
Education	-0.145	0.055	6.941**
Security	-0.478	0.082	33.636**
Compensation	0.177	0.085	4.352*
Constant	1.936	0.460	17.722
Nagelkerke R^2	0.152		
-2 log likelihood	512.701		
% correct prediction	63.8		

Notes: one and two asterisks (* / **) denote significance at the 5% and 1% levels respectively.

Education attainment exhibited strong predictive effects. The negative relationship suggests that the better educated respondents were more likely to have flood cover. All other socio-economic variables were not significant. These include household income, suggesting that wealthier households were no more likely to take out flood insurance.

Perceived importance of flood insurance in providing financial security was negatively associated with non-insurance, suggesting that those respondents who did not affirm the role of insurance were less likely to insure. It is important to note that expectation for disaster relief was positively related to non-insurance. Those who considered government compensation for flood victims to be a substitute to self-financed insurance were less likely to insure. This provides initial evidence on the pathological tendency of 'moral hazard' as discussed below. A caveat is that this variable managed to demonstrate marginal significance.

Surprisingly, the risk perception variable did not reach significance. A closer look at the data indicated a positive relationship: those who perceived the level of flood risk they faced as low, medium, high or extreme were more likely to have flood cover than those who felt negligible threats from floods. However, the claim for positive relationship is qualified by the low statistical significance ($\chi^2 = 4.652$, $p > 0.05$).Evidence did not permit the conclusion that perceived risk was a strong predictor. The way in which these observations appear at odds with the traditional view is discussed in the following section.

DISCUSSION

This research has found a moderately higher proportion (43.8%) of households reported to have no flood cover than that of another recent study (38%) recorded in Queensland (Tooth, 2012, p. 29). Differences in data collection methods could explain the discrepancy. In Tooth (2012), the non-insurance rate reported is based on contents insurance. Also, the respondents included residents of metropolitan and regional Queensland, and they completed the survey through the internet, which is a source of selection bias.

According to the official website of the Insurance Council of Australia (ICA) (http://www.insurancecouncil.com.au), Queensland has 76.6% of the current insurance policies with an active flood cover. This is much higher than the insurance coverage rate of 56.2% recorded by the present survey. The discrepancy is attributed to methodological variations, such as the inclusion of non-policy holders in our estimation but not in the ICA report[§], the use of the number of households as the basis of estimation rather than the number of policies, and the geographical focus on South East Queensland. The varied ways in which 'flood' is defined or understood by insurers and consumers (Australian Government, 2011) may also contribute to the discrepancy.

Behavioural economists argue that decision to adopt flood insurance is primarily influenced by risk attitude and thus non-insurance arises from misconception of flood risk (Kunreuther, 1996).The logical inference that risk aware individuals tend to purchase flood insurance finds limited empirical support from the present research. South East Queensland is known for its vulnerability to flooding, but ironically its residents, including those who

[§] The ICA estimate is calculated based on the number of *active* policies. Yet not all Queenslanders hold an active home insurance policy. Following the Productivity Commission's definition of 'non-insurance' (see Footnote 1), this study included non-policy holders in the estimation of non-insurance rate. This created a higher value of dominator and hence resulted in a lower insurance rate.

considered themselves to be at risk of flooding, were no more likely to buy flood cover than those did not. The behavioural importance of risk perception is not evident.

This observation challenges one of the ostensibly reasonable arguments often made by insurers as well as policy-makers. That is, 'householders who know they face a risk are inclined to purchase appropriate cover' (Insurance Council of Australia, 2011, p. 9). Instead, Handmer and Smith's (1989, p. 8) concern, raised more than 20 years ago, that 'it is unlikely that all those at risk would elect to take out flood insurance even if it was available' continues to have currency. Although it is always an imperative to raise risk awareness in the community by means of education (Australian Securities and Investment Commission, 2000; Carter, 2012; Godber, 2005; Yeo, 2002), it is unlikely to be a panacea to the enduring problem of non-insurance. Recognition of threat does not guarantee appropriate preventive actions.

Two other factors noted by previous studies, such as Roche et al. (2010) and Tooth (2012), are worth mentioning. Affordability does not offer a strong explanation for non-insurance, as indicated by the lack of significance of household income. Tooth (2012) claims that taxes on insurance make it less affordable and removal of those taxes could increase insurance rates. The industry strongly concurs with this view (Insurance Australia Group, 2011; Insurance Council of Australia, 2011; Powell, 2010). A counter-argument, however, is that those who could afford insurance are not necessarily inclined to purchase it. The wealth effect on insurance purchasing is complicated. As Baumann and Sims (1978) argue, wealthy households are more self-sufficient and can afford potential property damage, making flood insurance less attractive to them. Affordability is then unlikely to be a correlate. Furthermore, trust in insurers did not predict non-insurance. The household decision to adopt flood insurance was not discouraged by the lack of confidence in the industry.

The belief that flood insurance could shield the household against financial losses caused by flooding influenced the likelihood of non-insurance in a straightforward way. Affirmation of the potential benefits to one's own family ran in the same direction as the tendency to insure and proved to be the most powerful predictor in the model. This suggests that emphasizing the financial implications to the individual would be an effective strategy for motivating preventive actions at the household level, in contrast to a communal-interest framing which may lack personal relevance. Strengthening such beliefs may help reduce the incidence of non-insurance.

The positive relationship between expectation for disaster relief and non-insurance indicates the possibility of moral hazard and thus warrants regulatory attention. To those who see self-financed insurance and disaster relief as substitutes, knowledge of generous compensation for flood victims may strengthen the maladaptive belief that purchasing flood insurance is not necessary.

Declared formal commitments by the government in advance of a disaster require reconsideration in the context of the viability of commercial flood insurance(Mortimer, et al., 2011; Priest, et al., 2005). Of course, disaster relief is indispensable in the event of life-threatening natural catastrophes and it is politically untenable to remove it from emergency management plans. The marginal explanatory power of the moral hazard factor in the reported regression model does not justify a conservative shift in the direction of catastrophic finance. Other measures to minimize perverse incentives are needed.

CONCLUSION

Widespread non-insurance remains an unsettling challenge in Queensland. In this survey a significant minority of households reported to have no insurance against probable flood damage. The problem is not restricted to those who are unaware of the flood risks confronting them or those with relatively low incomes. The insured individuals are better educated and recognize the role of flood insurance in financially protecting the community. There is evidence for moral hazard although it is not conclusive. Attitudes toward insurance companies and insurance market are not found to be associated with non-insurance. To conclude, non-insurance is a function of the attitude toward the role of insurance, whereas the effects of risk perception are not evident.

There are two implications for flood risk management and insurance arrangement. First, although it is important to inform the community of the flood risks they face, risk education cannot guarantee substantial reduction in non-insurance rates. Reliance on the use of risk information tools, such as flood maps, is necessary but insufficient to improving the voluntary adoption of flood insurance. Second, prior knowledge of generous compensation after a disaster may create perverse incentives for non-insurance. Local administrators and flood managers should not only manage flood risk, but also the public expectations about institutional responses in the event of flooding.

Future research should explore how risk perception is shaped and identify how the attitudinal drivers of risk perception are related to the voluntary adoption of flood insurance. For example, risk perception may be mediated by social and cultural norms and understanding this mediating effect may be useful for characterizing the effect of risk perception. Other factors such as trust in institutions and scientific information may also be relevant and worth further exploring.

REFERENCES

Australian Securities and Investment Commission, (2000). Consumer understanding of flood insurance, Australian Securities and Investments Commission, Melbourne.

Australian Treasury, (2011). Natural Disaster Insurance Review, The Treasury, Australian Government, Canberra.

Baumann, D.D., Sims, J.H., (1978). 'Flood Insurance: Some Determinants of Adoption', *Economic Geography,* 54(3), 189-196.

Carter, R.A., (2012). 'Flood risk, insurance and emergency management in Australia', *Australian Journal of Emergency Management,* 27(2), 20-25.

Department of Climate Change, (2009). Climate Change Risks to Australia's Coast: A First Pass National Assessment, Department of Climate Change, Australian Government, Canberra.

Dowling, R.M., (2012). Report on the Effects of the January 2011 Flood on the Mangrove Communities Along the Brisbane River, Department of Environment and Resource Management, Queensland Government., Brisbane.

Godber, A., (2005). 'Urban floodplain land-use - acceptable risk?', *Australian Journal of Emergency Management,* 20(3), 22-26.

Handmer, J., (2002). 'Flood insurance: A backward glance and some current issues', in Residential flood insurance: the implications for floodplain management policy eds D. I. Smith and J. Handmer, Water Research Foundation of Australia, Canberra.

Handmer, J. (2008). 'Risk Creation, Bearing and Sharing on Australian Floodplains', *International Journal of Water Resources Development,* 24(4), 527-540.

Handmer, J., Smith, D.I., (1989). 'Flood insurance and relief in Australia: The background', in Flood insurance and relief in Australia, eds D. I. Smith and J. Handmer, Centre for Resource and Environmental Studies, Australian National University, Canberra.

House of Representatives Standing Committee on Social Policy and Legal Affairs, (2012). In the Wake of Disasters, Volume One: The operation of the insurance industry during disaster events, Commonwealth of Australia, Canberra.

Hung, H.C., (2009). 'The attitude towards flood insurance purchase when respondents' preferences are uncertain: a fuzzy approach', *Journal of Risk Research,* 12(2), 239-258.

Insurance Australia Group, (2011). Submission to Productivity Commission Inquiry into Regulatory and Policy Barriers to Effective Climate Change Adaptation, Insurance Australia Group.

Insurance Council of Australia, (2011). Insurance Council of Australia Response to 2011 Natural Disaster Insurance Review, Insurance Council of Australia, Sydney.

Kriesel, W., Landry, C., (2004). Participation in the National Flood Insurance Program: An Empirical Analysis for Coastal Properties', *Journal of Risk and Insurance,* 71(3), 405-420.

Kunreuther, H., (19960. Mitigating disaster losses through insurance, *Journal of Risk and Uncertainty,* 12(2), 171-187.

Lim, G.C., Chua, C.L., Claus, E., Nguyen, V. H., (2012). Review of the Australian Economy 2011–12: A Case of Déjà Vu', *Australian Economic Review,* 45(1), 1-13.

McDonald, J., (2010) Paying the price of adaptation: compensation for climate change impacts. In : Bonyhady, T., Macintosh A., McDonald, J. (Eds), Adaptation to climate change: law and policy, *The Federation Press,* Annandale, NSW.

Mortimer, E., Bergin, A. and Carter, R., (2011). Sharing risk: Financing Australia's disaster resilience, Australian Strategic Policy Institute, Canberra.

Phelan, L., Taplin, R., Henderson-Sellers, A., Albrecht, G., (2011). Ecological Viability or Liability? *Insurance System Responses to Climate Risk, Environmental Policy and Governance,* 21(2), 112-130.

Powell, S., (2010). An insurer's approach to management of climate risks. In : Bonyhady, T., Macintosh A., McDonald, J. (Eds), Adaptation to climate change: law and policy, *The Federation Press,* Annandale, NSW.

Priest, S.J., Clark, M. J. and Treby, E.J., (2005). Flood insurance: the challenge of the uninsured', *Area,* 37(3), 295-302.

Productivity Commission, (2013). Barriers to Effective Climate Change Adaptation. Report No. 59, Final Inquiry Report, Productivity Commission, Australian Government, Canberra.

Queensland Government, (2011). Climate Change: Adaptation for Queensland Issues Paper, Department of Environment and Resource Management, Queensland Government, Brisbane.

Queensland Government, (2012) Queensland Floods Commission of Inquiry - Final Report Queensland Government, Brisbane.

Reserve Bank of Australia, (2011). Financial Stability Review March 2011, Reserve Bank of Australia, Canberra.

Roche, K.M., McAneney, K.J. and Van Den Honert, R.C., (2010). 'Policy options for managing flood insurance', *Environmental Hazards,* 9(4), 369-378.

Smith, D. I., (2002). The what, where and how much of flooding in Australia. In: Smith I., Handmer, J. (Eds), Residential flood insurance: the implications for floodplain management policy Water Research Foundation of Australia, Canberra.

Smith, D. I. and Handmer, J., (2002) A summary of flood insurance issues and options. In: Smith I., Handmer, J.(Eds), Residential flood insurance: the implications for floodplain management policy Water Research Foundation of Australia, Canberra.

Thomas, M., King, D., Keogh, D. U., Apan, A., Mushtaq, S., (2011). Resilience to climate change impacts: a review of flood mitigation policy in Queensland, Australia, *Australian Journal of Emergency Management,* 26(1), 8-17.

Tooth, R., (2012). Australian Household Insurance: Understanding and Affordability, Sapere Research Group, Sydney.

Yeo, S. W., (2002) 'Flooding in Australia: A Review of Events in 1998', *Natural Hazards,* 25(2), 177-191.

In: Flooding
Editors: M. R. Motsholapheko and D. L. Kgathi

ISBN: 978-1-63117-607-4

Chapter 7

HABITUS: HOW CULTURAL VALUES SHAPE LOCAL COMMUNITIES' PERCEPTIONS ABOUT FLOODS IN THE OKAVANGO DELTA OF BOTSWANA

O. D. Kolawole*

Okavango Research Institute, University of Botswana

ABSTRACT

Floods and flooding events are of central interest in the studies bordering on the Okavango Delta ecosystems, the sustainability of which depends on regular water flow. Nonetheless, as beneficial as flood pulses might be to the river basin and the riparian communities in and around it, extreme flooding events continue to impact on rural livelihood systems and people's well-being in the area. This chapter employs the concept of Pierre Bourdieu's [1930-2002] habitus and the use of qualitative data (obtained through key informant interviews) to analyse and explain how cultural values shape people's perceptions and how they respond to natural phenomena (such as floods), which impinge on their living conditions. Through the application of Kurt Lewin's [1890-1947] field theory and 3-step model of planned change, and in partial combination with Bourdieu's field, the discourse offers insights on how scheduled change agencies could better understand the social forces that perpetuate undesired and desired behaviours of individuals comprising their clientele systems and how this understanding could enhance the application of appropriate planned change program for achieving behavioural change in the periods of emergency triggered by water inundation.

Keywords: community, perception, floods, riparian, culture, livelihoods, field, social change, behaviour

* Author e-mail: tkolawole@ori.ub.bw; toyin_kolawole@yahoo.com

INTRODUCTION

The incidents of natural hazards and disasters are not new to mankind. While a natural hazard connotes any hydrological, geophysical and atmospheric event (e.g., flood, earthquake, tsunami, cyclone, drought, etc.) which can constitute potential loss or harm to the people and environment, a natural disaster is an extremely hazardous event, which could result in disruptions and, loss of lives and properties in any human settlement or community thereby leaving the people at the mercy of external aid or assistance (Benson and Twigg, 2007). Extreme weather events experienced across the globe in the recent times continue to pose a threat to humankind particularly the vulnerable and poor people. Admittedly, these extreme events are connected with the global climate change and continental Africa is among other regions expected to bear the burden resulting from these alterations (see Schaeffer *et al.*, 2013). Thus sub-Saharan Africa will bear the brunt of the devastating effect of inclement weather conditions induced by changes in weather patterns. Southern Africa, where Botswana is situated, is not an exception. In the recent times, climate variability and extremes have been felt in the form of extreme flooding and droughts in the region. While the 'normal' rainfall in southern Africa ranges from 50 mm to over 1000 mm, recent changes in weather events have resulted in erratic rainfall leading to either extreme floods or severe droughts (Chenje and Johnson, 1994; World Meteorological Organisation [WMO], 2000). Amongst others, the disruption of farming activities by a combination of severe dry spells and floods from 1999-2001 impacted negatively on food production, making many countries in the region to grapple with severe food shortages. For instance, Zambia, Namibia and Botswana experienced significant declines in coarse grain production as a result of these extreme climate events (United Nations Food and Agricultural Organisation [FAO], 2001).

Through a critical discourse analysis, this chapter employs the concept of habitus and field theory to unearth the sociological and cultural dynamics that influence people's predisposition towards certain 'deviant' behaviours within a social milieu. Using case study design, it goes further to analyse the influence of culture on people's perception about and predisposition to natural hazards like floods. The chapter then employs Kurt Lewin's 3-stage model of planned change to explain and suggest how community people's perceptions and predisposition could be positively changed in the events of natural disasters like extreme flooding within and around the Okavango Delta of Botswana.

The chapter begins by introducing the reader to the ecological dynamics of the Okavango Delta in north-western Botswana, and the people who live in the area.

Okavango Delta, People and Floods

The Okavango Delta, which is one of the largest freshwater environments in southern Africa and one of the largest inland deltas in the world, covering an area of about 15000 square kilometres of pristine and natural feature of diverse flora and fauna (Zambezi Safari Travels [ZST], 2012; see also, Mendelsohn *et al.*, 2010). The uniqueness and natural splendour of the Delta informed the unanimous decision of international environmental experts to officially declare it as one of the *7 Natural Wonders of Africa* in Arusha, Tanzania on 11th February 2013 (Seven Natural Wonders of Africa [SNWA], 2003). Not existing in isolation, the Okavango Delta derives its water from the Cuito and Cubango Rivers which flow from the central highlands of Angola. Triggered by huge subtropical storms, the rivers flow through Namibia's Caprivi Strip* as Kavango River and then finally entering northern Botswana at Mohembo (see Figure 1). Each year, an average of 9.3 million cubic meters of water empties into the Delta (Mendelsohn *et al.*, 2010: p42) and drains away into the Kalahari wastes of the south plains through lagoons and channels (see ZST, 2012). The cyclical nature of water flow and the attendant seasonal floods (see Mendelsohn *et al.*, 2010: p40) to a high degree influence the livelihood systems and pattern of the inhabitants living in and around the Okavango Delta. Riparian communities and people found along the river banks of the Delta are largely engaged in fishing, arable farming and livestock husbandry. Thus flood recession farming traditionally known as *Molapo* farming is popular amongst the people. It is instructive to note that these riparian community people - through the community-based natural resource management (CBNRM) framework - also engage in tourism activities such as tour guiding and traditional boat (*mokoro*) rides, offering tourists to explore the river channels of the Okavango Delta.

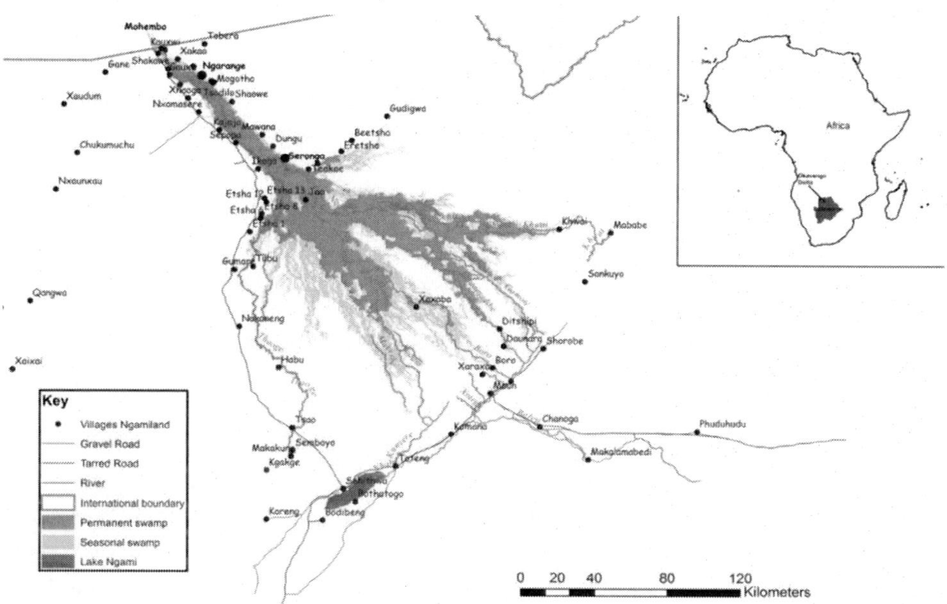

Figure 7.1. Okavango Delta map showing riparian communities exposed to flood situations (Courtesy: Masego Dhliwayo and Anastacia Makati, GIS Laboratory, Okavango Research Institute, University of Botswana, Maun.

The strong attachment, which these communities have with the Delta and its resources, dates back to many years of interactions with the aquatic ecosystem. Ironically and perhaps influenced by the effect of climate change and variability, most floodplains within the Delta did not experience flood pulse for almost three decades up to 2009. The dryness experienced

* In what appears like a move to obliterate Namibia's German colonial history, the government on the 8th August 2013 renamed the Caprivi Strip (a 450km^2 area popular for its tropical rivers and wildlife) as the Zambezi Region, after the river that forms the northern border with Angola. (see http://www.news24.com/Africa/News/Namibia-renames-Caprivi-Strip-20130808)

by the people within this period of long dry spells significantly altered the land use patterns such that people began to erect landed properties in the floodplains where ordinarily they would not have done so in the period of regular water flow. Nonetheless, there was a major turn of event in the periods 2004 and 2009-2010 during which the flood forcefully came back.

In 2009 and 2010, riparian communities in northern Botswana such as Nata, Gweta, Seronga, Tubu, Ikoga, Nxamasere, Etsha 13, Mohembo East and Nxaraga experienced extreme and destructive flooding induced by the Okavango River (Mosate, 2010) leading to loss of properties (International Federation of Red Cross [IFRC], 2009) and agricultural harvests (Motsholapheko, 2011). During the period, an estimated 620 families or 3798 persons had been affected in Nata, Gweta, Seronga, Ikoga, Tubu, Nxamasere, Etsha 6 and Etsha 13 where about 600 huts and 76 modern houses were damaged by the floods. Apparently warned about the impending flood by relevant scheduled agencies, some community people, because of their cultural values and spiritual attachments to the land, did not evacuate from their properties already situated along the path of the flood (Kolawole, 2012). Somehow, certain community people's awareness about the impending danger, which the flood might constitute, seemed not to be in congruence with their perceptions about the magnitude of the problem and the need to find alternative means of livelihoods. Their acculturation with flood events and how they manage them as well as the need to stay close to their means of livelihoods, which the risk-averting messages conveyed to them did not cater for, informed the need to hold fast to the known rather than the unknown. Although acknowledged that associated water borne diseases and reptiles could pose a threat to their well-being (see for instance, Ditlhakeng et al., 2012), community people continue to brave adverse ecological situations in their search for survival. This will be revisited shortly.

Theoretical Underpinning

Habitus and People's Pre-Disposition to Floods

The thrust of this chapter finds its relevance in the concept of habitus, which originally was an idea mooted by Aristotle. Although many scholars approach the concept from different perspectives, the proposition as to how and why people in a social group are culturally entrenched in a particular way of ordering their lives, and then predispose themselves to certain behaviours in record history makes Pierre Bourdieu's work more appealing in the context of this writing. Unlike the body habitus (medically known as physique), which explains the phenotypic and genotypic constitutions of an individual's bodily make-up and how he or she by these becomes naturally predisposed to some form of disease(s) (American Heritage Dictionary of the English Language [AHDEL], 2003), Bourdieu's habitus simply explains an individual's make-up as engendered by his or her social environment and certain value orientations associated with it. Habitus typifies the cultural underpinnings and structures embedded in people's bodies and minds. Thus, the notion of how people perceive and read meanings to the physical, natural and socio-cultural phenomena around them is particularly rooted in Bourdieu's (1977) habitus in relation to its dependency on history and human memory. For instance, a learned behaviour or belief within a given cultural setting is perpetuated through the socialization process within a social structure long after the reason for the behaviour or belief has been forgotten or can no longer be recalled by the emerging new generations in that given culture. Simply defined, habitus depicts 'the way society becomes deposited in persons in the form of lasting dispositions, or trained capacities and structured propensities to think, feel and act in determinant ways, which then guide them' (Wacquant, 2005: 316) in their predisposition and creative responses to overcoming the challenges associated with their external environment. It is the objectification of social structure in the individual's subjectivity; the 'internalization of externality and the externalization of internality'. In other words, habitus is created through a social, rather than an individual process leading to certain social patterns that are long lasting and transferrable from one social context to another but which also shifts in relation to specific social contexts and over a period of time. In Mauss' (1934) own term, habitus is an embodiment of material and non-material dimensions of culture peculiar to certain individuals, groups, societies, etc., which are most apparent in the 'learned habits, bodily skills, styles, tastes, and other non-discursive knowledge that might be said to "go without saying" for a specific group…' and this is perceived to operate below the level of 'rational ideology'.

Habitus in relation to the stages of its development, takes two distinctive forms – the primary and secondary habitus (Wacquant, 2013; Bourdieu and Passeron, 1977 [1970]: pp42-46). In his own terms, Wacquant (2013) conceives primary and secondary habitus as 'generic' and 'specific', respectively. While primary habitus is basically the 'experiential' mode by which people directly acquire and learn about their societal way of life through non-formalized, socialization process, secondary habitus is the 'didactic' or formal educational mode through which all other multiple habitus are acquired (Wacquant, 2013). It is thus impossible for an individual to acquire any variety of specific habitus without first acquiring the generic habitus either peculiar to his or her own family of orientation or social grouping. As such, the primary habitus is the foundation for the secondary habitus. In Loïc Wacquant's own words, '[e]very agent has a primary (generic) habitus, which is both springboard and matrix for the subsequent acquisition of a multiplicity of specific habitus' (Wacquant, 2013). Indeed, it is safe to infer that the unique experiences and the personality acquired by an individual within a given social setting are as a result of the habitus associated with the society in which the individual grows and develops. Habitus is thus an embodiment of a social process rather than that of an individual's. It is about the unfolding of a group in relation to how members perceive and interpret the daily occurrences within their culture. Perhaps cognitive mapping and validation play a significant role in the way people perpetuate certain belief systems within a given society. How people bury their dead, how they relate with one another on a day-to-day basis, how they get attached to certain symbolic and cultural traits, the reasons as to why people conduct themselves in certain ways in a given social system and most of which can no longer be explained, are a form of habitus. For instance, whenever rural dwellers are obliged to explain the reason(s) behind their responses and reactions to certain phenomena (whether natural or social), we are often informed: 'That is how our forefathers did it, and we cannot afford to deviate from the norm. But then, habitus 'is not fixed or permanent, and can be changed under unexpected situations or over a long historical period' (Navarro 2006: p16). Indeed habitus is not rigid but 'fully amenable' and malleable to new experiences (Wacquant, 2013; Bourdieu, 2000 [1997]: p161) acquired by the individual through some form of organized, educational processes. Many years before reflexivity became popular amongst scholars, Pierre Bourdieu had proposed a 'reflexive sociology'. Thus Bourdieu was of the opinion that people need to recognise their biases, beliefs and assumptions in a bid to make sense out what happens around them (Navarro, 2006). In other words, in-depth and careful reflections will allow people see why 'rationality' should take pre-eminence over 'irrational' decisions and actions. I will return to this in the following section.

In the present context, the analysis of how community people get permanently attached to their Ancestral land has some bearing on their conceptualization of the spiritual dimension of property acquisition, ownership and management. The reason as to why some people find it difficult to detach from such personal properties in the advent of potentially catastrophic natural events (like flood, cyclone, etc.) is essentially the thrust of this chapter. Although they may have received warning signals, the presumed notion of the need to save properties without an iota of reflections about the jeopardy this might cause precious life and people's well-being may probably find explanation in how people's perceptions are shaped by their habitus. Better still, the need to hang on and not let go the existing livelihood systems and the difficulty one might experience in embracing change and living in a totally different socio-ecological environment may further explain the why people find it difficult to relocate whenever the need to do so arises. Although flood causes disruptions to human life (Meyer and Bendsen, 2003; Motsholapheko et al., 2011), some of the reasons why the affected communities do not seem to heed warning signals in the advent of flood have been established by literature. These include the discrepancy in the way institutions and communities perceive risks (Magole and Thapelo, 2005), poor communication infrastructures and logistics (Ditlhakeng et al., 2012), low level of education within rural communities (VanderPost, 2010), which limits job opportunities in the formal sector and possibly elsewhere (Ditlhakeng et al., 2012). Indeed, the ecological peculiarity of Botswana is a major factor influencing how people perceive and respond to water issues. Situated in the midst of the semi-arid Kalahari region, the entire country is parched and receives little rainfall during any given raining season. This single most important factor makes people to attach great importance to water! The Batswana associate rain with some form of blessing. This is locally referred to as *Pula*. A riparian community elder provided a balanced viewpoint on the positive and negative impacts of floods on his people during a key informant interview session in December 2013. He remarked thus:

> 'Floods are perceived as a form of blessing because they bring life to our people. As we rely more on the river for our daily livelihood activities, we get more food such as fish and water lilies, which serve as vegetables. During the floods, our mode of transportation changes and becomes less expensive when we have high floods; we travel to other villages mostly by canoes as opposed to the not-easy-to-come-by vehicular transportation. But since 2009, however, we have been receiving above normal floods. Although flood does not affect our houses, it impacts negatively on our daily chores and vehicular movement and social service provision (like health care)...We end up travelling long distances to seek medical attention. As animal movements are restricted in the events of such flood episodes, livestock starvation and mortality are not uncommon. Floods also induce human suffering and ailments such as diarrhoea, skin diseases and malaria, and other fevers. When we experience flood, our youths by their nature quickly relocate to other villages because they are always on the move. This leaves the elderly people in the community helpless as the old people rely mostly on the assistance provided by the younger ones.' (Mr. Galethuse: Daonara community).

While community people acknowledge that floods could constitute a natural hazard to their well-being, their perceptions about the benefits that water availability confers on community sustenance and livelihood systems are crucial and more important than anything else. It is understandable then that local people who regularly experience long drought spell and whose major source of livelihood depends mainly on water would thoroughly embrace flood howsoever the magnitude would be in any given climatic event. Rather than perceive flood (be it extreme or otherwise) as a destructive force, community people see it as a part of the '…biodiversity production system and a source of livelihood' (Magole and Thapelo, 2005) on which traditional flood recession farming system depends. Most certainly then, habitus, which predisposes an individual to avoid situations or conditions to which he or she is not naturally adapted, appears to be a strong influence in community peoples' conservative tendency to stick to what they have. While relevant government institutions and non-governmental organizations (NGOs) were 'jittery' about the implications of the impending flood on riparian communities (such as Tubu) in 2004, many local people who would be primarily affected appeared unperturbed (Magole and Thapelo, 2005) and would readily surmise that 'flood is a phenomenon to which we are accustomed'. This viewpoint was further confirmed in a key informant interview exercise conducted in Tubu community in December 2013 during which the village *Kgosi*[†] shed more light on how community people perceive floods. He remarked:

> 'Floods are a blessing to our people because they do not adversely affect our houses. Instead, floods provide good opportunity for us to use canoe as the mode of transportation. More importantly, a flooding event brings with it nutrients to our fields which allows for good farming and food production. Besides, we are able to collect wild fruits from the river banks when there is enough water in the channels. Floods also stimulate the eventual growth of fresh fodders on which our livestock easily graze. And the aftermath of floods is the rapid growth of reeds and other tall grasses [meant for roof thatching], which we harvest for sale. Floods also prevent fire outbreaks in the surroundings. We find it more convenient to stay put here during floods because this is the only life we know – we grow up here and are adapted to the environment; thus we build our houses on relatively higher elevations. Yes, we have challenges but the only threat to life is crocodiles, which we sometimes encounter; they do attack our livestock.' (*Kgosi* Motshidiemang of Tubu community)

Clearly, two things emerged from the opinions of the village chief. First, community people perceive that the benefits derived from floods far outweigh the adversities they (floods) bring. Two, riparian community people are used to aquatic environments and are seemingly able to make internal and self-directed adjustments whenever the need arises, thus seeing relocation as unnecessary. Nxamasere community provides a good case study. While some residents of the village, who supposedly were hard hit by the flood events of 2009-2010 were relocated by the Land Board (Mosate, 2010), many members of the community did not relocate, possibly because they felt they were not seriously affected or they perceived that they could brave the consequences of adverse ecological conditions. Perhaps the reasons why certain community people are unwilling to relocate in the event of adverse floods (as provided by another village elder in Daonara community) will suffice:

[†] In Botswana, the village headman or chief is traditionally referred to as *Kgosi*; when more than one, they are known as *Dikgosi*. In a way, the village headman or chief thus represents the voice of the community people.

> 'Most elderly people do not have the penchant for relocating to certain designated places when we experience floods because they are not used to living in canvass tents [locally known as Dixhibi] provided for them by [the] aid or government agency; they would rather prefer to live under their own traditional houses. They prefer to remain in their settlement to plough on their own fields rather than start a new farming life in an unknown terrain where there are likely to be more predators and crop raiders. Besides, staying here enables us to readily have access to our own local food, which we prefer over and above the modern ones that are common in the urban centres.' (Mr. Galethuse: Daonara community)

The community elder's viewpoints above portray a strong boundary maintenance, which local people perpetuate in order to sustain their identity (see Loomis and Beegle, 1950) and stick to the known rather than the unknown. Thus the inability of local, elderly people to dwell in any other seemingly inferior forms of abode other than their traditional architectures even in 'times of emergency', and their inability to eat any other forms of food other than their local delicacies is an indication of a strong affinity for one's own traditions.

Elsewhere in the mid Okavango Delta, community people in Jao Flats[‡] once discountenanced government's directive on the need for them to relocate but rather saw the move as a form of control by the powers that be (Magole and Thapelo, 2005). Partly buttressing this claim, one key informant who is a Village Development Committee (VDC) chairman provided a vivid explanation on how community people perceive and respond to flooding events thus:

> 'We do not see floods as a threat to us. Jao community is what we have known as our home; we have no other place to go! Our Ancestors lived here and we have all the attachment to this area. The only unpleasant experience we normally have is mainly in crocodile attacks on livestock – the beasts live along the river channels and they rarely go beyond that point. And our children are always instructed on how to avoid them. *Contrary to the viewpoints of mass media people on the surrounding water surge, we do not perceive increased water inflows as constituting any hazards as they (the news reporters) would make people believe - they do not provide correct and adequate information about disaster management* [Emphasis mine]. We are a people who are so much conversant with water; we are born in a water environment and raised therein. As we are accustomed to canoe mode of transportation, moving around becomes easier for us when we have enough water around here.' (Mr. Motswai: Jao Flats community)

As earlier indicated, strong attachment to land and properties, and local people's viewpoints about spiritual and Ancestral connections are some of the reasons why relocation becomes almost impossible in times of crises! Indeed, the VDC chairman's viewpoints above further support other key informants' opinions on their perceptions about floods. More importantly, there is a dissonance in the way warning messages are conveyed to community people by [the] relief or government agencies and how these messages are construed or decoded by the recipients of the information. It may then imply that cognitive, psychomotor (conative) and affective[§] components of habitus (see for instance, Wacquant, 2013) find relevance in how institutions and change agents convey early warning messages and how

[‡] Jao Flats is a small island community located in the heart of the Okavango Delta (see Figure 1).

[§] The three concepts of cognitive, psychomotor and affective domains are originally rooted in Bloom's taxonomy of educational or learning objectives (see Bloom, 1994).

community people in-turn perceive and read meanings to them. First, the way in which each party perceives and assigns meaning to flood and its implications (i.e., cognition) will play a crucial role as to whether or not both the change agents and community people would reach a consensus in addressing the flood problems at hand. Second, the dexterity with which the change agent employs his or her mental skill to physically drive the change process (i.e., psychomotor) as observable to and understood by the clientele system plays a significant role in bringing about social change. Third, the extent to which the change agent shows enthusiasm and readiness to invest his or her energy in undertaking the process of change [for which s/he is employed] with the sincere intent of seeking community people's well-being has a positive correlation with the measure of success achieved; one could probably predict the outcomes of events from the on-set where and when a change program is implemented perfunctorily by the change agent(s). Regardless of the impact, which their uncompromising stance might have on community socio-economic and physical well-being, groups who see themselves as vulnerable and powerless tend to engage in muscle-flexing by exhibiting passive resistance to government order, particularly in situations where they feel that the dividends of good governance have failed to reach them meaningfully.

Culture and People's Perception About Flooding Events

Culture is a way of life of a people, which is learned, shared and transmitted from one generation to the other through the socialization process. It is encompasses shared ideas, norms, values and belief systems of any social grouping (see Ekong, 2003: pp23, 390). How people perceive their social space and how they read meanings to certain events (be they social, physical or natural) are influenced by their value orientations embedded within the cultural milieu in which they are situated. Traditionalism, fatalism and immediate gratification are some of the negative orientations that could impinge on community people's objective reflexivity in a bid to bring about improvement in their lives. But then, people naturally have the ability to: take control of their own lives through a reference value (a mental image/construction of the desired state); have a perceptual function (the ability to observe the status quo or existing state); devise a mechanism for making comparisons (the ability to compare the existing state and the desired state for differences); act to bring the existing state closer to the desired state (L. Smith, 1997; see also Kolawole, 2002). More often than not, how community people perceive a problem situation and not just a mere awareness of it would determine the kind of response or solution they are likely to provide in surmounting the problem by which they are confronted (see Kolawole, 2002). And individuals or a group's perceptions about the problem at hand are a function of the culture in which they find themselves.

Indeed, flooding events and how people perceive them are central to the argument in the chapter. In common parlance, flood portends danger or disaster (Magole and Thapelo, 2005) and has been seen as the commonest of all natural disasters, and that which brings both benefits and losses more than any other natural hazards thinkable (Smith, 1991: p259-260). While floods now constitute a perennial danger to both lowland and upland communities due to climatic change being experienced globally, communities in wetland areas such as those in the Okavango Delta perceive the situation differently where '...floods are critical for maintaining and restoring many of the important services provided to humans by wetland ecosystems' (Magole and Thapelo, 2005). As a result of the peculiar ecological system in which they find themselves, local farmers and community people in the Delta naturally rely on flood water for their livelihoods as rainfall scarcity and extreme dryness are a commonplace in the Kalahari region of southern Africa. In the Delta for instance, such benefits include the creation of a better habitat for biodiversity in both aquatic and terrestrial flora and fauna, improved agricultural lands through alluvial deposits in the floodplains (Wisner *et al.*, 2004: p205) where local farmers engage in traditional flood recession (*molapo*) farming. The VDC chairman interviewed in Jao community corroborated this claim, saying that:

> '…this place is the only suitable location where we can thoroughly engage in our traditional (*Molapo*) farming activities; the soils are naturally replenished with nutrients by the annual inflow, and we engage in other sources of livelihoods such as basketry. Inflow of water enables us to have access to wild fruits and palm fruits (*Mokolwane*), which serve as food for us. In addition, there are only few animals that destroy our crops unlike what obtains in far away upland settlements.' (Mr. Motswai: Jao Flats community)

Clearly, traditional agriculture is a component of the cultural base of rural communities. The specifics associated with a particular type of farming system and other livelihood strategies are apparent in the cultural traits and patterns prevalent in any local community. The riparian community people's value orientations as reflected in their traditional approach to farming deserve special interests. *Molapo* farming (which is a form of traditional agriculture whereby local farmers rely on flood recession to cultivate and tend their crops) is an age-long practice amongst the Batswana farmers found within and around the Okavango Delta. To this people, the cyclical flood pulses experienced as an annual event [during normal hydrological periods] in the area is perceived as a 'form of blessing' because of its importance to farming and other livelihood activities such as fishing, hunting and even tourism. Corroborating the importance of floods to tourism employment, a community elder affirmed that *[w]e derive other benefits such as employment provided by Community Trusts, which are not easy to come by when we go to other places in the hinterland* (Mr. Galethuse: Daonara community).

Regardless of the awareness about the unpredictability and extent of the flood pulses, which could engender loss of lives and properties in extreme cases (Magole and Thapelo, 2005); local people continue to brave flooding events and their associated ecological adversities whenever they happen. As earlier noted, the people's perceptions about the happenings around them and how they ascribe meanings to them far supersede the awareness of the events or phenomena. For instance, some southern African community people do not see soil erosion as a problem until it fully becomes a gully because 'erosion-induced and localized soil types were believed to be so created by God' (Kolawole, 2002; 2006; Cartier and Graaff, 1998)! Interestingly, what is often defined and classified as an urgent and emergency problem by development agencies may not necessarily be so seen by local communities (see also Jongmans, 1981; Jungerius, 1986). Contrary to the notion that local people are often risk-averse and are not willing to venture into the unknown because of their low level of education and lack of exposure (see Kolawole, 2002), it is indeed paradoxical to note that many riparian community people on the one hand continue to defy ecological challenges to sustain their livelihood systems and uphold their attachment to land but on the other hand, and in an intricate manner, also exhibit a measure of risk aversion by not willing to find help elsewhere whenever the need arises to relocate to some other 'safe haven'. Where possible, the need to identify and implement behavioural change program through context-specific educational initiatives, which seek to enhance the knowledge, skills, and attitudes of the clientele system is indeed an imperative.

To move away from the status quo and ensure a meaningful social action program that enhances ruralites' objective reflexivity in the midst of chaos and uncertainty, there is need to shift and change old paradigms and come up with new frameworks in the process of alleviating the vulnerable conditions of the rural poor in the wetland plains of the Okavango Delta. The following section therefore explores the ways and means of effecting behavioural change in the development process in a peculiar social-ecological context.

FLOODS AND PLANNED CHANGE: REFLECTIONS ON THE OKAVANGO DELTA

In this section, I endeavour to explore practical approaches to behavioural change as it relates to the response of communities to floods and flooding events within the Okavango Delta. I wholeheartedly acknowledge Pierre Bourdieu's own proposition, which places emphasis on bridging the gap between objectivism and subjectivism through the relationship between habitus and field[**] in social science research practice (see for instance, Wacquant, 2005; Navarro, 2006). Nonetheless, rather than use Bourdieu's field theory to fully complement his concept of habitus in this chapter, I have chosen a slightly innovative and integrative route to accomplish the task. The reasons are not far-fetched. First, Bourdieu's field theory is mainly interested in the social activities that take place within a 'social room' comprising many fields. The social room is the larger society, which is the equivalent of a social system. The field is likened to the sub-systems operating within a given social system (e.g., farming, academic, religious and business communities). Within each field exists a set of 'rules of access' that govern people's (agents')[††] actions, preferences, social relationships and practices, and how they perceive themselves and those outside their own field. Those sets of rules - acceptable standards, sanctions, norms, etc. - are what Bourdieu referred to as 'doxa' (Bourdieu, 1984: p141). Individuals in a field (community, in this case) must play to the rule of the game if they are to gain better positions and access the resources within that field. The fields are stratified based on social statuses resulting from people's ability to acquire certain capitals (social, economic, cultural or symbolic) (Bourdieu, 1986). Thus Bourdieu's field theory is about how individual agents struggle and compete to gain new positions and access resources (see for instance, Gaventa, 2003) by playing to the rules of the game associated with their own field. In sum, Pierre Bourdieu's field theory is central to power struggle and

[**] Pierre Bourdieu's field connotes 'a setting in which [individual] agents and their social positions are located'. The interaction between the individual's agent's habitus and capital (social, economic and cultural) and the prevailing sanctions or rules in the field determines his or her position in that given field. Fields interact with each other in a hierarchical fashion, depicted in power and class relations (Bourdieu, 1984). Thus a field is a social arena where individuals struggle and compete to acquire necessary resources for survival.

[††] The concept of agent in Bourdieu's field connotes the individual person operating and competing within a social arena

domination; it emphasizes how people imbibe societal ideals, preferences, choices, actions, social differences and hierarchies and power relations amongst members and those of other fields or communities outside theirs. It suggests the consciousness of one's own place and 'self exclusion' (Bourdieu, 1986: p471). It is nothing more than positions, tastes and power. Second, Bourdieu's theory does not offer any veritable pathway as to how to minimize the gap between desired behaviours and undesired behaviours of individuals within a social field as found in Kurt Lewin's propositions. While Bourdieu's work offers useful insights on how individuals predispose themselves to the elemental frameworks and machinations of their immediate environment, and by so doing enables the change agent to better appreciate and understand, to a considerable degree, the dynamics of any social grouping, it however falls short of concrete analytical schema of the fundamentals of human behaviour and the cogent pathway pointing to the practicality of resolving problems associated with undesired behaviour in a planned change program. As Gaventa (2003) puts it, '…Bourdieu concentrated [largely] on analyses of the media and academia, however, so it would be difficult to apply these insights directly to the development 'field''.

Relevant and most appropriate to this discourse, therefore, is the treatise of Kurt Lewin (1890-1947) on planned change. Planned change simply implies 'a direct and human intervention in the shaping and direction of change toward some predefined goals' (Ekong, 2003: p262), which as it were, is meant to deliberately alleviate people's miserable conditions. It is designed to counteract the effect of any unplanned or accidental change associated with natural disasters caused by extreme floods, cyclones, earthquakes, locust invasion, etc. which in themselves do not allow people to adequately plan before they occur. As such, planned change is a form of social action carefully geared towards achieving some desirable development goals within a given social milieu. For instance, the need for government to devise and implement an educational and awareness cum relocation program is an initiative deliberately targeting flood-prone communities in the advent of extreme flooding. Nonetheless, to successfully achieve this goal would mean that there is need to properly understand the dynamics of social groupings and behaviours of individuals in a community setting in relation to why and how they would respond to any external stimuli in any given [unpleasant] scenario.

That said, field theory enabled Kurt Lewin and his research team to identify and better understand the forces that perpetuate undesired social behaviours and those which reinforce desired ones in their attempt to fashion out strategies to either strengthen or weaken those attributes (as the case may be) so as to bring about the desired behaviours in people (M. Lewin, 1998). In other words, Kurt Lewin's desire to understand individual behaviour informed the development of a field theory, which was originally applied in physics. Nonetheless, he later used it 'mainly as a method for analysing and changing group behaviour' (Burnes, 2007).

Rooted in gestalt[‡‡] psychology that emerged in Germany in the early 20th Century (Köhler, 1967), Lewin's theory posits that an individual behaviour is a product of the totality of the co-existing and interdependent [social] forces, which impact on an individual or a

‡‡ Psychologists conceive a gestalt as 'a perceptual pattern or configuration that is the construct of the individual mind. It is a coherent whole that has specific properties that can neither be derived from the individual elements nor be considered merely as the sum of them' (Kadar and Shaw, 2000 in Burnes and Cooke, 2012).

group within a 'life space' (otherwise referred to as psychological, social or force field[§§]) where the behaviour takes place (Lewin, 1942). As Lewin believed that an individual's behaviour is a product of his or her environment and how he or she reads meaning to external stimuli, taking into consideration the intricate nature and totality of that individual's psychological or perceptual environment while attempting to construct the person's life space will enable the change agent to understand, predict and change the person's behaviour (Lewin 1943a). And to discern the individual or group's life space as conceived by Lewin himself, there must be an understanding of the person or group's verbal reports and how they perceive or make sense out of their own situations (Deutsch, 1968: p416). It is therefore not enough to base an individual or a group behaviour merely on the external stimuli that interfere with their state of equilibrium; the individual or group's subjective perceptions of the forces that impinge on them go a long way in explaining why they behave in the way they do. A combination of flood events and how community people perceive those occurrences in relation to their well-being is vividly captured in various locations within the Delta where village chiefs/elders and key informants voiced their opinions on how they perceived these natural happenings. The understanding that flood pulses are a life-giving, natural phenomenon without which riparian community's livelihood systems cannot be sustained explains why people would not necessarily see their (flood pulses) occurrences as a threat to life in the first place. While community people individually agree that floods could cause certain disequilibrium in the well-being of the community, their positive outlooks about the advantages conferred by flood events far outweigh their negative perceptions about this natural event. Whereas outsiders may have considered floods as a risk, it is 'a form of blessings' to those who are directly affected by them. Again, riparian community people's habitus - as evoked by the preference for one's own indigenous food and architecture, and a strong attachment to the land and Ancestors as well as the premium placed on certain peculiar agricultural systems (e.g., flood recession farming) - brings to the fore the reasons why people respond to flood in the way they do.

Thus the four major elements that comprise Lewin's planned approach to change include field theory, group dynamics, action research and the three-step model of change (Burnes 2004). But in this chapter, attention is focused on Lewin's field theory, the 3-step model and their relevance in the implementation of social change programs in rural communities. Indeed planned social change primarily originated from Kurt Lewin's (1939; 1940) field theory. Among others, and of much interest to this chapter, are two characteristics of the field theory attributable to Kurt Lewin. These attributes include the dynamic and psychological approaches which underline his field theory wherein he argued on the basis of the dynamic nature of the social forces in a life space. To him, a recognizable form of equilibrium is maintained in a social life constantly undergoing a dynamic process. Ultimately, alterations which appear as undercurrents regularly occur in a social life although they may go unnoticed because of their tendency to maintain a stable condition. He coined this as a 'quasi-stationary equilibrium' (Lewin, 1947) and posited that an alteration in and from one 'quasi-stationary equilibrium' to another is ultimately engendered by an alteration in the psychological forces in a life space (Lewin, 1943b). In the context of this analysis, community people and elders quickly find a way of adjusting to their new realities when flood events bring certain

[§§] For Kurt Lewin, a field is an environment in which an activity takes place. By implication then, a social field is where a social activity takes place.

unpleasant shifts. Thus the inability to travel by roads as a result of water inundation does engender an immediate shift from the 'not-easy-to come-by' road transportation to water travels through the use of canoe, which of course is perceived as cheaper and mostly preferred by the people. Ensuring that life must continue, the elderly people naturally would devise measures to circumvent any form of instability created by youth massive relocation to other communities during the floods.

From the viewpoint of psychological approach, Lewin argued that how individuals or a group perceive(s) their reality at a given point in time must be the observer's primary focus rather than attempting to construct it from his or her own 'objective' viewpoint (Burnes and Cooke, 2012). Therefore, the change agent must be cognizant of the fact that two different people who experienced the same phenomenon (e.g., a flooding event in this case) may have perceived the event differently (see Rock and Palmer, 1990). It is instructive to note that what is troublesome for one group may not necessarily be so for another (see Mendelsohn, *et al.* 2010: 110). Based on the information obtained from key informants, flood occurrences to some people, are a natural means for preventing bush fire and a period during which livestock fodders are readily available. To others, it is a trying time when animal movements are restricted, thus engendering livestock starvation, disease and loss! Indeed, the dissonance existing between an individual or a group's realities and those of others (outside their immediate environment) is most apparent in the viewpoints of the VDC chairman in Jao who commented that the mass media and government people's perceptions about flood is totally at variance with those of community people. While the former perceives flood as constituting a serious danger to life and property, it is a blessing to the latter - those who are accustomed to the natural event.

That said, Kurt Lewin's proposition - that individual behaviour is a function of the field or life space at the time it occurs but not in the past or future, and Pierre Bourdieu's habitus emphasis on history and human memory seem to have headed on a collision course. While Lewin's emphasis is on how individual or group behaviour is affected by the 'here and now' (Deutsch, 1968), Bourdieu's viewpoint on the same hinges mainly on what has happened in the past – an historical event. But seen differently, I reckon that an individual's behaviour [within a group setting] is most likely influenced and more strongly impacted by a combination of both past and current events. In other words, experiences of the past and the existing ones tend to reinforce individuals' perceptions about life events and the prevailing phenomena around them, particularly so if those unique experiences do not tend towards dissonance.

It is only when individuals and groups are assisted to better comprehend and reflect on the social forces that encroach and impact on their lives that behavioural change could be achieved (Lewin, 1942). In line with Bourdieu's reflexivity on the need for people to think objectively about the happenings around them, Lewin's approach to planned change, which supposedly would create a platform that will enable individuals understand and reorganize their perceptions of the world around them (Burnes, 2007; Lewin, 1942) forms the basis for devising a participatory strategy for ensuring a pro-active social action and change. To a considerable degree, it appears that habitus and field theory are indeed not mutually exclusive; they complement each other in finding solutions to behavioural change. While Bourdieu's habitus seeks to elaborate on the individual constituents and how certain habits or ways of life are formed, replicated or perpetuated within a social space, Lewin's field theory places emphasis on the identification of factors or forces that perpetuate undesired behaviours and the factors that need to be strengthened or weakened in order to bring about the desired behaviours in individuals or groups within a given social field (M. Lewin, 1998). But then, achieving a change in behaviour would mean that other elements of Lewin's planned approach to change (i.e., action research and 3-step model) are fully deployed in the process of implementing a planned change program. In other words, while field theory and group dynamics are primarily concerned with the intricacies involved in the formation of social groupings and how they are stimulated and sustained, action research and the 3-step model are interested in how behaviours of individuals and groups are changed (Burnes and Cooke, 2012). Thus, both habitus and field theory serve as the foundation or structure on which the superstructures (i.e., any planned change programs) are built.

Over all, a holistic action-oriented research and development program that incorporate analytical techniques and issues that address the need to understand the forces that impinge on people, and how multiple realities of individuals and groups could be reduced to a single reality will help in bringing about behavioural change amongst community people.

Kurt Lewin's 3-Step Model of Planned Change

Although pummelled by certain criticisms that Lewin's Planned approach is too simplistic, mechanistic and has failed to recognize politics and power, and the incessant conflicts associated with human organizations, among others (see for instance, Dawson, 1994; Hatch, 1997; etc.), but all of which have been objectively countered by some scholars (e.g., Burnes, 2004), Kurt Lewin's (1947), the three-step model of planned change, which is noted for its persistent influence on creating and managing change (see Hendry, 1996: 624) finds relevance here and can therefore be appropriately invoked. The three phases or stages comprise (1) unfreezing; (2) transitioning; and (3) refreezing. Through the application of force field analysis, the first stage occurs where and when the social forces that are striving to maintain the 'status quo' and subjective 'mindsets' are identified and dismantled or broken down by making use of appropriate communication channels. In the process, all attempts are carefully and painstakingly made to ensure that stakeholders or the clientele system see the enormity of the problem(s) at hand. To prepare the stage for transitioning and for people to completely move away from the status quo, Schein (1996) cited in Burnes (2004) had suggested 3 processes of (i) invalidating the status quo; (ii) inducting guilt or survival anxiety into the people; and (iii) creating psychological safety in their mind. For people to accept change and then transit into the new order, they must have concrete reason(s) why they must do so and '...those [directly and ultimately] concerned have to feel safe from loss and humiliation' (Burnes. 2004) that may arise due to the social change they suddenly undergo. Otherwise the unfreezing phase will not occur in the first place! Openness and honesty are required to get the job done. Transitioning, which Burnes (2004) also refers to as 'moving', involves the development of new sets of attitudes, behaviours, values, etc. through a well organized educational program and other relevant strategies to minimize the chaos that might ensue in the process of adapting to new change. The key to this second stage is good communication, negotiation, involvement and empowerment; community people relocated from flood prone areas would need some forms of adaptable socio-economic and cultural empowerment to enable them cope with the new change. The last stage, which is the refreezing phase, involves the crystallization and reinforcement of the new values, attitudes, behaviours, etc. This process involves the development of appropriate instruments to reward compliance to the new change, recognition and celebration of success, sanctions wherever and whenever possible, and sustained training and education. Otherwise there is the possibility for some people to return to the *status quo ante*!

Given this context, those attributes, which tend to subjectively predispose people to some seemingly 'illogical' and 'irrational' thinking [in response to emergency situations like flooding events] need to be identified and unfrozen, while desirable and pro-active behaviours could be frozen over a considerable period of time. Where and when by consensus people's multiple realities become a single reality with respect to a common cause or goal, there is an emergence of a single understanding amongst the people or within any organization about the need to bring about meaningful and successful changes in their co-existence (see for instance, Boje and Rosile, 2010). Through context-specific and empathetic educational programs, it is thus assumed that the riparian communities of the Okavango Delta will be better equipped to understand the dynamics and implications of adverse effects of floods on their psychosocial and economic well-being.

CONCLUSION

The first section of this chapter provided a brief synopsis on the dynamics of the Okavango Delta river flow, and the socio-economic and cultural activities of the people living in the area. While the second section used the concept of habitus to explain why riparian community people in the Delta respond to floods in the way they do, sections 3-4 employed a critical analysis of Kurt Lewin's field theory [and partially complemented by Bourdieu's field] and the 3-step model to highlight key issues in understanding people's behaviour and how to apply these in bringing about community behavioural change in the period of natural disasters.

The lack of understanding of designated change agencies or government departments and non-governmental organizations (NGOs) as to why and how people behave in the way they do in times of extreme flooding and emergencies may have, in the past, engendered their (i.e., government and NGOs) failings and inability to convince certain community people to relocate from disaster-prone areas. Many community people were supposedly 'laid back' in their response to early warning messages provided them by relevant agencies possibly because their multiple realities about floods do not result in a single reality or possibly because their realities are not in agreement with those of the change agencies. Either way, there is a clientele-change agency reality dissonance. Most certainly, problems arise in the 'casting' of any local individuals' viewpoints about flood if there is a wide gap or 'distance' between their experiential knowledge (i.e., primary habitus) of the phenomenon and the kind of educational messages (i.e., secondary habitus) passed to them by the change agency. Assuredly, '[t]he greater that distance, the more difficult the traineeship' (Wacquant, 2013) on how to make local people perceive and manage flood disasters. Viewed from another perspective, the hierarchical positions and power relations amongst individuals within and between social arenas [fields] as enunciated by Bourdieu's field theory may explain why community people predispose themselves in the way they do in their interactions with government agencies in extreme circumstances like flood disasters. The notion of 'us' and 'them' as depicted in Bourdieu's field constitutes a major barrier to any planned change program in situations where community people see external agencies (be they governmental or non-governmental) as 'exploitative', 'apathetic' and 'untrustworthy'. Many development programs did not succeed because change agencies fail to properly understand the social fields in which their clienteles operate. Not only does this affect behavioural change, it has implications for a whole gamut of rural development programs in general. Behavioural change in itself is a learning process in which the acquisition of new knowledge, attitudes and skills play key roles in making people to change their 'perceptions, insights, outlooks, expectations [and] thought patterns' or gain new ones (French and Bell, 1990) about the events around them.

The findings show that although they admit that there are many ills associated with flooding, riparian community people believe that flood events confer on them many advantages (as they relate to community livelihoods and survival) over and above all the disadvantages of flooding. This disposition brings to the fore the significant effect of societal deposition in individuals. To a considerable extent, it indeed shows the influence of habitus on any people's behaviour and judgment about the happenings in their immediate surroundings. Whether we like it or not, rural people's cultural and environmental knowledge, which they have acquired over time, continues to shape their viewpoints and perceptions about themselves and the outside world. It is thus the primary onus of the change agency to devise strategies for effectively invalidating the people's perceptions about the need to adhere to certain customs and preference for livelihoods instead of personal safety and survival. It is also the duty of the change agent to find a way of motivating and convincing community people (through a well coordinated campaign and awareness programme) that bravery lies in one's ability to see dangers in adverse [ecological] situations and avoid them while at the same time activating in them a survival anxiety; a life fully lived is that which is replete with hope anyway. The change agency would also need to create in the people's mind some measure of safety psychology if only to convince endangered people to hid the voice of reasons in times of impending natural catastrophes.

Invariably, a positive result is achieved where subjectivity gives way to objectivity. But even so and admittedly, too, people are dynamic and complex; and so does development itself. Thus the solution for now may not necessarily be ideal for the future (see Mendelsohn, *et al.*, 2010). The understanding that people's perceptions about a particular phenomenon vary even though they have both experienced it in a similar way could however help the change agent to appreciate people and group's behaviours in a relatively complex human organization and society. This will enable him or her identify the best strategy or approach to adopt in implementing a meaningful planned change program. There can be no futile efforts where the change agency and agent labour hard and strive diligently to close the wide gap that may have existed between the clientele's generic and specific habitus (on disaster management). Indeed, a multi-sectoral and disciplinary approach is needed to enable stakeholders see and analyze the problem situation through different development lenses. In so doing, an eclectic approach rather than a simple paradigm may have been more appropriate in addressing social-ecological problems and people's relocation in times of extreme floods.

ACKNOWLEDGMENTS

I am most grateful to the Institute's Field Technical Officer - Mr. Ronald Mothobi – for his timely assistance in ensuring a good surrender value in qualitative data gathering during this write-up. I also acknowledge US NSF-START for the fund awarded in conducting a climate change research, which afforded me the opportunity to have a first-hand interaction with community people on pertinent social-ecological issues, and from which some of the data used in this write-up emanated.

REFERENCES

AHDEL, (2003). *The American Heritage Dictionary of the English Language* (4th Ed) Houghton Mifflin Company.

Benson, C., Twigg, J., (2007). Tools for mainstreaming disaster risk reduction: guidance notes for development organisations, International Federation of Red Cross and Red Crescent Societies/the ProVention Consortium, January 2007, Geneva: ProVention Consortium Secretariat. On-line document: *http://www.preventionweb.net/files/ 1066_toolsformainstreamingDRR.pdf* (Accessed 3 December 2013).

Bloom, B.S., (1994). Reflections on the development and use of the taxonomy. In Anderson, L.W. and Sosniak, L.A. (eds.) Bloom's taxonomy: A forty-year retrospective. Chicago, IL: *Chicago National Society for the Study of Education.*

Boje, D., Rosile, G.A., (2010). *Lewin's action research meets Aristotle on the storytelling field: maybe it's a distinctive niche for CTU.* Colorado Springs, CO: Colorado Technical University.

Bourdieu P. (2000 [1997]). Pascalian meditations. Cambridge: Polity Press.

Bourdieu, P. (1986). The forms of capital. In J. Richardson (Ed.) *Handbook of theory and research for the sociology of education*, New York: Greenwood Press, pp. 241-258.

Bourdieu, P. (1984). *Distinction: a social critique of the judgement of taste.* London: Routledge.

Bourdieu, P. (1980). *The logic of practice.* Stanford: Stanford University Press.

Bourdieu, P. (1977). *Outline of a theory of practice.* Cambridge: Cambridge University Press.

Bourdieu P., Passeron J-C. (1977 [1970]). *Reproduction in education, society and culture.* London: Sage.

Bourdieu, P., Wacquant L.J.D. (1992). *An invitation to reflexive sociology.* Chicago: The University of Chicago Press.

Burnes, B. (2007). Kurt Lewin and the Harwood studies: the foundations of OD. *Journal of applied behavioral science,* 43: 213–231.

Burnes, B. (2004). Kurt Lewin and the planned approach to change: a re-appraisal. *Journal of management studies*, 41: 977–1002.

Burnes, B., Cooke B. (2012). Kurt Lewin's field theory: a review and re-evaluation. International journal of management reviews, pp. 1-18, DOI: 10.1111/j.1468-2370.2012.00348.x On-line document: *http://onlinelibrary. wiley.com/doi/10.1111/ j.1468-2370.2012.00348.x/pdf* (Accessed 28 August 2013).

Cartier van Dissel, S., de Graaff J. (1998): Differences between Farmers and Scientists in the Perception of Soil Erosion: a South African case study. *Indigenous knowledge and development monitor* 6(3): 8-9.

Chenje, M., Johnson, P. (Eds.). 1994). *State of the environment in southern Africa.* SADC/IUCN/SARDC, Maseru and Harare.

Dawson, P. (1994). *Organizational change: a processual approach.* London: Paul Chapman Publishing.

Deutsch, M. (1968). Field theory in social psychology. In: Lindzey, G. and Aronson, E. (eds), *The Handbook of Social Psychology*, Vol. 1, 2nd edition Reading, MA: Addison – Wesley, pp. 412–487.

Ditlhakeng, P., Mthombeni, N., Sommer, C. (2012). *Analysis of factors influencing the low adoption of flood risk information in flood prone communities in the Okavango Delta, Botswana, Winter course research report,* Okavango Research Institute, University of Botswana, pp. 9-75.

Ekong, E.E. (2003): *Rural sociology: an introdution and analysis of rural Nigeria,* Dove Educational Publishers, Uyo, Nigeria.

FAO (2001). 17 Countries are facing exceptional food emergencies in sub-Saharan Africa - FAO concerned about deteriorating food situation in Sudan, Somalia and Zimbabwe. Press Release 01/48, Nairobi/Rome: Food and Agriculture Organization of the United Nations. On-line document: *http://www.fao.org/waicent/ois/press_ne/presseng/2001/pren0148.htm* (Accessed 14 December 2013).

French, W.L., Bell, C.H. (1990). *Organization development, 4th edition.* Englewood Cliffs, NJ: Prentice Hall.

Gaventa, J. (2003). Power after Lukes: An overview of theories of power since Lukes and their application to development, Brighton: Institute of Development Studies, p. 9. On-line document: *http://www.powercube.net/wp-content/uploads/2009/11/power_after_lukes.pdf* (Accessed 30 September 2013).

Hatch, M. J. (1997). Organization theory: modern, symbolic and postmodern perspectives. *Oxford: Oxford University Press.*

Hendry, C. (1996). Understanding and creating whole organizational change through learning theory. *Human relations*, 48(5): 621–641.

IFRC (2009). *Botswana floods. Botswana, 30 June 2009.* Geneva: International Federation of Red Cross.

Jongmans, D.G. (1981). The farmers of Khroumirie in relation to their natural surroundings. In: *Nature as refuge, explorations in cultural ecology*, 8. Delft: Delftse Universitaire Pers. pp. 1-4.

Jungerius, P. D. (1986). Perception and use of physical environment in peasant societies. *Geographical papers*, Department of Geography, University of Reading, p. 2.

Kadar, E.E., Shaw, R.E. (2000). Toward an ecological field theory of perceptual control of locomotion. *Ecological psychology*, 12: 141–180.

Köhler, W. (1967). Gestalt psychology. *Psychological research, 31: XVIII–XXX.*

Kolawole, T. (2012). *The plot lies on a floodplain. Botswana Guardian*, 4 May, p.16.

Kolawole, O.D. (2006). *Local farmers approach to soil conservation: lessons from Nigeria. Indlinga: African journal of indigenous knowledge systems* 5(1): 77.

Kolawole, O.D. (2002). *Factors associated with the utilization of indigenous knowledge systems for soil fertility conservation by farmers in Ekiti state,* Nigeria. PhD Thesis,

Department of Agricultural Extension and Rural Sociology, Obafemi Awolowo University, Ile-Ife, Nigeria, p.33.

Lewin, K. (1947). Frontiers in group dynamics. In: Cartwright, D. (ed.) (1952). Field theory in social science: Selected theoretical papers by Kurt Lewin. *London: Social Science Paperbacks*, pp. 188–237.

Lewin, K. (1943a). Defining the 'Field at a given time'. In: Cartwright, D. (ed.), (1952). Field theory in social science: selected theoretical papers by Kurt Lewin. *London: Social Science Paperbacks*, pp. 43–59.

Lewin, K. (1943b). Psychological ecology. In: Cartwright, D. (ed.) (1952). Field theory in social science: selected theoretical papers by Kurt Lewin. *London: Social Science Paperbacks*, pp. 17–187.

Lewin, K. (1942). Field theory and learning. In: Cartwright, D. (ed.), (1952). Field theory in social science: selected theoretical papers by Kurt Lewin. *London: Social Science Paperbacks,* pp. 60–86.

Lewin, K. (1940). Formalization and progress in psychology. In: Cartwright, D. (ed.) (1952). Field theory in social science: selected theoretical papers by Kurt Lewin. *London: Social Science Paperbacks*, pp. 1–28.

Lewin, K. (1939). Field theory and experiment in social psychology: concepts and methods. *American journal of sociology*, 44: 868–896.

Lewin, M. (1998). Kurt Lewin: his psychology and a daughter's recollections. In: Kimble, G.A. and Wertheimer, M. (eds.), *Portraits of Pioneers in Psychology,* Vol. 3. Mahwah, NJ: Lawrence Erlbaum Associates, 105–120.

Loomis, C.P., Beegle, J.A. (1950). *Rural social systems: a textbook in rural sociology and anthropology*. Englewood Cliffs, NJ: Prentice-Hall.

Magole, L., Thapelo, K. (2005). The impact of extreme flooding of the Okavango River on the Livelihood of the *molapo* farming community of Tubu Village, Ngamiland, Botswana. Botswana Notes and Records, 37: 1-13. On-line document: *http://www.ubrisa.ub.bw:8080/bitstream/handle/10311/949/Magole%202005%20Impact%20of.pdf?sequence=1* (Accessed 27 August 2013).

Mauss, M. (1934). Les Techniques du corps, Journal de psychologie 32(3-4). Reprinted in Mauss, *Sociologie et anthropologie*, 1936, Paris: PUF.

Mendelsohn, J.M., Vanderpost, C., Ramberg, L., Murray-Hudson, M., Wolski, P., Mosepele, K., (2010). *Okavango Delta: floods of life*. Windhoek: RAISON

Meyer, T., Bendsen, H. (2003). The dynamics of the land use systems in Ngamiland: changing livelihood options and strategies, in *Environmental Monitoring of Tropical and Subtropical wetlands*. Bernard, T., Mosepele, K. and Ramberg, L. (eds.) Okavango Report Series No. I; Maun, Botswana: HOORC, 278 -307

Mosate, M. (2010). An update on the floods situation. Botswana, Government Portal, Botswana, 12 July 2010. On-line document: *http://www.gov.bw/en/PrintingVersion/?printid=6770* (Accessed 3 June 2012).

Motsholapheko, M. R., Kgathi, D. L., Vanderpost, C. (2011). Rural livelihoods and household adaptation to extreme flooding in the Okavango Delta, Botswana. *Physics and Chemistry of the Earth, Parts A/B/C,* 36(14-15), 984-995.

Navarro, Z. (2006). In search of cultural interpretation of power, *IDS Bulletin* 37(6): 11-22.

Rock, I., Palmer, S. (1990). The legacy of Gestalt psychology. *Scientific America,* 263(6): 48–61.

Schaeffer, M., Baarsch, F., Adams, S., de Bruin, K., De Marez, L., Freitas, S., Hof, A., Hare, B. (2013): Africa's adaptation gap: climate-change impacts, adaptation challenges and costs for Africa, Technical Report, United Nations Environment Program (UNEP). On-line document: *http://www.unep.org/pdf/AfricaAdapatationGapreport.pdf* (Accessed 3 December 2013).

Schein, E.H. (1996). Kurt Lewin's change theory in the field and in the classroom: notes towards a model of management learning. *Systems practice,* 9(1): 27–47.

Smith, L.C. (1997): Introducing MIKS: a methodology for the utilization of indigenous knowledge. *Indigenous knowledge and development monitor* 5(2): 17-18.

Smith K. (1991). *Environmental hazards: assessing risks and reducing disasters.* London, Routledge.

Seven Natural Wonders of Africa [SNWA], (2013). Okavango Delta, The Seven Natural Wonders. On-line document: *http://sevennaturalwonders.org/africa/okavango-delta/; http://naturalwondersofafrica.com/okavango-delta/* (Accessed 1 October 2013).

Vanderpost, C. (2010). *Socio-economic change in Ngamiland.* Fact sheet 9. Maun, Okavango Research Institute.

Wacquant, L. (2013). Homines in extremis: what fighting scholars teach us about habitus. *Body and society*, pp. 1-15. DOI: 10.1177/1357034X13501348.

Wacquant, L. (2005). *Habitus. International encyclopedia of economic sociology.* J. Becket and Z. Milan. London: Routledge.

Wisner, B., Blaikie, P., Cannon, T., Davis, I. (2004*). At risk: natural hazards, vulnerability and disasters.* (2nd edition), London: Routledge.

World Meteorological Organisation [WMO]. (2000). WMO Statement on the Status of the Global Climate in 2000.*WMO, Geneva*, 19 December 2000.

Zambezi Safari Travels [ZST], (2012). The source of the Okavango Delta. The Zambezi Safari and Travels Co. Ltd. On-line document: *http://www. zambezi.co.uk/safari/ botswana/okavango.html* (Accessed 3 March 2012).

In: Flooding
Editors: M. R. Motsholapheko and D. L. Kgathi
ISBN: 978-1-63117-607-4

Chapter 8

RURAL LIVELIHOODS AND HOUSEHOLD ADAPTATION TO EXTREME FLOODING IN THE OKAVANGO DELTA, BOTSWANA*

M. R. Motsholapheko†, D. L. Kgathi and C. Vanderpost
University of Botswana, Okavango Research Institute, Botswana

ABSTRACT

Adaptation to flooding is now widely adopted as an appropriate policy option since flood mitigation measures largely exceed the capability of most developing countries. In wetlands, such as the Okavango Delta, adaptation is more appropriate as these systems serve as natural flood control mechanisms. The Okavango Delta system is subject to annual variability in flooding with extreme floods resulting in adverse impacts on rural livelihoods. This study therefore seeks to improve the general understanding of rural household livelihood adaptation to extreme flooding in the Okavango Delta. Specific objectives are: 1) to assess household access to forms of capital necessary for enhanced capacity to adapt, 2) to assess the impacts of extreme flooding on household livelihoods, and 3) to identify and assess household livelihood responses to extreme flooding. The study uses the sustainable livelihood and the socio-ecological frameworks to analyse the livelihood patterns and resilience to extreme flooding. Results from a survey of 623 households in five villages, key informants, focus group discussions and review of literature, indicate that access to natural capital was generally high, but low for financial, physical, human and social capital. Households mainly relied on farm-based livelihood activities, some non-farm activities, limited rural trade and public transfers. In 2004 and 2009, extreme flooding resulted in livelihood disruptions in the study areas. The main impacts included crop damage, household displacement, destruction of household property, livestock drowning and mud-trapping, the destruction of public infrastructure and disruption of services. The main household coping strategies were labour switching to other livelihood activities, temporary relocation to less affected areas, use of canoes for

* Also in *Journal Physics and Chemistry of the Earth* © Reproduced and modified with permission from ELSEVIER Limited, The Boulevard, Langford Lane Kidlington, Oxford, UK

† Corresponding author. Address: University of Botswana, Okavango Research Institute, P/Bag 285 Maun, Botswana. Telephone: +267 686 1833. Email: rmoseki@ori.ub.bw

early harvesting or evacuation and government assistance, particularly for the most vulnerable households. Household adaptive strategies included livelihood diversification, long-term mobility and training in non-agricultural skills. The study concludes that household capacity to adapt to extreme flooding in the study villages largely depends on access to natural capital. This is threatened by population growth, land use changes, policy shifts, upstream developments, global economic changes and flood variations due to climate variability and change.

Keywords: adaptation, flooding, livelihoods, Okavango Delta

INTRODUCTION

Current trends in flood response show that there is a shift from mitigation to adaptation in order to lessen the impacts of flooding on human activities and livelihoods (Schipper and Burton, 2009). This is partly due to realisation that flood control measures have limitations due to changes brought about by climate variability and change (Few, 2003 Niang *et al.*, 2007; Schipper and Burton, 2009). Additionally, many flood control measures are beyond the capacity and capability of many developing countries, particularly when there is a high frequency of high floods. The low capacity to adapt to the impacts of flooding and other shocks in developing countries is associated with their low levels of human, financial, physical and natural capital, as well as weak institutional capacity or poor governance (Niang *et al.*, 2007; Vanderpost, 2007; Eriksen *et al.*, 2008; Leary *et al.*, 2008). At the micro-economic level, this results in high vulnerability to shocks among households, particularly those that depend on natural resource-based livelihoods.

In the Okavango Delta, Botswana, flooding is a major biophysical event that influences human-environment system interactions. It enhances ecosystem performance by supplying water and nutrients through sediment transfer to support the rich biodiversity as well as household livelihoods (McCarthy *et al.*, 2003). These positive dimensions contrast with the negative impacts of extreme flooding that occurred during the 1970s, 1980s and again from 2009. The occurrence of extreme floods causes human life disruptions. The floods disrupt cultivation in floodplains, destroy property leading to displacement of households, interrupt water reticulation systems, and curtail transport systems affecting the transfer of goods and services (Bendsen and Meyer, 2003; Magole and Thapelo, 2005; Malala, 2009).

Rural households have over long periods adapted in various ways to variable flooding levels. However, socio-economic changes such as population growth, economic transformation, alterations in household social fabric, land use policy and administrative changes have negatively affected past adaptive strategies (Wilk and Kgathi, 2007; Dube and Sekhwela, 2008).

The impacts of extreme flooding on rural livelihoods in the Okavango Delta have been documented (Bendsen and Meyer, 2003; Magole and Thapelo, 2005). However, the relative importance of the various adaptive strategies, the causes of low adaptive capacity and the vulnerability of households are still not well understood. Past incidents of this shock in various parts of the country have been reported in the 1970s and 1990s (Tsheko, 2003). Little is known about the long-term impacts of floods on household livelihoods. According to Central Statistics Office (2009: p3), "known reliable records on floods" may be recent dating from 1995. This may be attributed to the generally rare occurrence of extreme flooding and the relatively new institutional preparedness for this shock in Botswana. There is a specific need to identify the causes of low adaptive capacity and high vulnerability among rural African households, especially those that become victims of the impacts of extreme flooding and other livelihood shocks. This will help build strategies for strengthening their capacity to adapt.

The purpose of this paper is to improve the general understanding of rural household livelihood adaptation to extreme flooding in the Okavango Delta by: 1) assessing household access to forms of capital necessary for enhanced capacity to adapt 2) assessing the impacts of extreme flooding on household livelihoods in the Okavango Delta, and 3) identifying and assessing household responses to extreme flooding.

Conceptual Framework, Study Area and Methods

Conceptual Framework

The study uses the sustainable livelihood framework, to understand how livelihoods are constructed, and the socio-ecological framework, to explain the livelihood vulnerability context in the Okavango Delta. The two frameworks are generally similar, save for the way they contextualise a livelihood in relation to shocks.

According to the sustainable livelihood framework, a household constructs a livelihood by combining five forms of assets being natural, human, social, physical and financial capital (Ellis, 2000). Access to or ownership of capital is modified by social relations, institutions and organisations which are endogenous to the household as they are within its control (Ashley and Carney, 1999; Ellis 2000). Household access to capital is also impacted on by shocks, trends and seasonality regarded as exogenous factors as they are beyond household control (Ellis, 2000; Scoones, 1998). This process results in livelihood strategies with effect on livelihood security and environmental sustainability (Ellis, 2000).

The framework has some limitations in that it does not cater for changes over time (Ellis, 2000). Furthermore, the vulnerability context in the framework does not consider the full context of shocks, trends and seasonality. It does not clearly show how negative livelihood outcomes can also exacerbate the impacts of shocks. It also obfuscates the role of the private sector within the construct of organisations that mediate access to forms of capital. To overcome some of these limitations, the sustainable livelihood framework may be modified or supplemented using other frameworks. For instance, in a study of livelihoods in Mali, Brock (1999) overcame the limitation on change over time, by choosing study sites where some historical data were available. Odero (2006) emphasizes the inclusion of information as a form of capital. In this study, the socio-ecological framework was used to explain the vulnerability context within which extreme flooding occurred.

The socio-ecological framework explains human development and adaptation in the context of the coupled human-environment interactions (Berkes and Folke, 1998). It analyses rural development in terms of clusters of elements (or characteristics) and interactions among clusters that finally lead to a sustainable society or community (Arntzen, 1989; Berkes and Folke, 1998). In this study, the interactions of interest are in the four elements of, 1) the

Okavango Delta ecosystem services provided to humans, 2) people and technology 3) local knowledge, and 4) property rights institutions. These elements were found to fit well within the constructs of natural and human forms of capital as well as institutions as earlier explained in the sustainable livelihood framework. They define the type of natural resources households exploit and their ability to adapt, build resilience to shocks and reduce vulnerability (Berkes and Folke, 1998; Folke, 2001). The main contribution derived from the socio-ecological framework is in locating the livelihood system within the broader context of human-environment interactions. Adaptation analysis should examine the following aspects: the shock itself, the impacted system characteristics and responses (Smithers and Smit, 2009). Understanding the interactions in the coupled human-environment system can help us appreciate that adaptation occurs within a complex set of factors inherent in the system and that climatic changes have ecological and socio-economic impacts (Turner II *et al.,* 2003).

The nature of the shock constitutes risk (the hazard itself) or biophysical vulnerability, while the household conditions define sensitivity or social vulnerability (Brooks, 2003). Intergovernmental Panel on Climate Change (2001: p95) defines vulnerability as "... a function of the character, magnitude, and rate of climate variation to which a system is exposed, its sensitivity, and its adaptive capacity." Shocks can be defined by their frequency of occurrence, speed, and their spatial and temporal dimensions, all of which constitute biophysical vulnerability (Brooks, 2003). Depending on their spatial coverage, shocks may be described as covariate or idiosyncratic. Covariate shocks affect the entire system while idiosyncratic shocks are specific within a system. In the Okavango Delta context, extreme flooding can be described as a covariate shock of slow onset and short duration. It affects the entire livelihood system, occurs slowly across the wetland and lasts several months of the year. It is exogenous to the household and, when it occurs, it acts as a stimulus that warrants response in coping, ultimately leading to adaptation (Ellis, 2000; Smithers and Smit, 2009). Household socio-economic, cultural and political characteristics define the sensitivity or social vulnerability, while adaptive capacity comprises the ownership of or access to all forms of capital (Hahn *et al.* 2009). Government interventions in the form of relief assistance and social welfare make up an important element for building coping and adaptive capacity.

Study Area

The study was undertaken in the Okavango Delta situated in the North West District of Botswana. The district had a population of 122 024 people in 2001, with a growth rate of 2.8% in the inter-censal decade 1991 to 2001 (Central Statistics Office, 2002). is found in The district capital Maun accounted for one third of this population, the rest being in rural settlements. The North West District is culturally diverse with at least five ethnic groups with distinct land and water resource management practices.

In the last two decades, the district had growth in economic activities, partly attributed to the development of tourism (Mbaiwa, 2002). However, there is rampant poverty among 49.7% of households, high unemployment (estimated at 30.7%) and HIV/AIDS prevalence (27.3%) rate (Ministry of Finance and Development Planning, 2007). Since the 1980s the North West District had major changes in land use and policy related to the control of animal diseases, wildlife management and tourism related activities.

The Okavango Delta, is an alluvial fan with an aerial coverage of 12 000 km^2 and a generally low gradient (McCarthy et al., 2003; Wolski and Murray-Hudson, 2006). The Delta (Figure 8-1), receives annual floods the size and coverage of which depend, mostly on rainfall in the catchment area of the Cuito and Cubango rivers in central Angola.

Mean annual rainfall in the Okavango Basin varies from 876 mm and 983 mm in the Cuito and Cubango rivers, respectively, to 450 mm in the Delta (McCarthy *et al.*, 2000; Wolski and Murray-Hudson, 2008). It is generally low and erratic contributing about 25% of all the water, and droughts are common (Anderson *et al.*, 2003). As a natural system, the Delta has inter-annual variability with periodic oscillations of wet and dry cycles (McCarthy *et al.*, 2000; Wolski and Murray-Hudson, 2008). Rainfall variations in the catchment have occurred in both the Cuito and Cubango, with gradual decline since the late 1970s (Gumbricht *et al.,* 2004; Wolski & Murray-Hudson, 2008). Consequently, there were periods of very low inflow which contributed to desiccation of some parts of the Delta in the early 1990s to mid-2000s. The Delta is currently in the wet phase with more flooding observed in 2009 and 2010 (Wolski, 2011 personal communication).

Time series data from early 1930s (Figure 8.2), indicate that high inflows with possible extreme flooding have periodically occurred. The highest inflow was in 1968 measuring 16 000 Mm^3 at Mohembo hydrological station (Harry Openheimer Okavango Research Centre [HOORC], 2010).

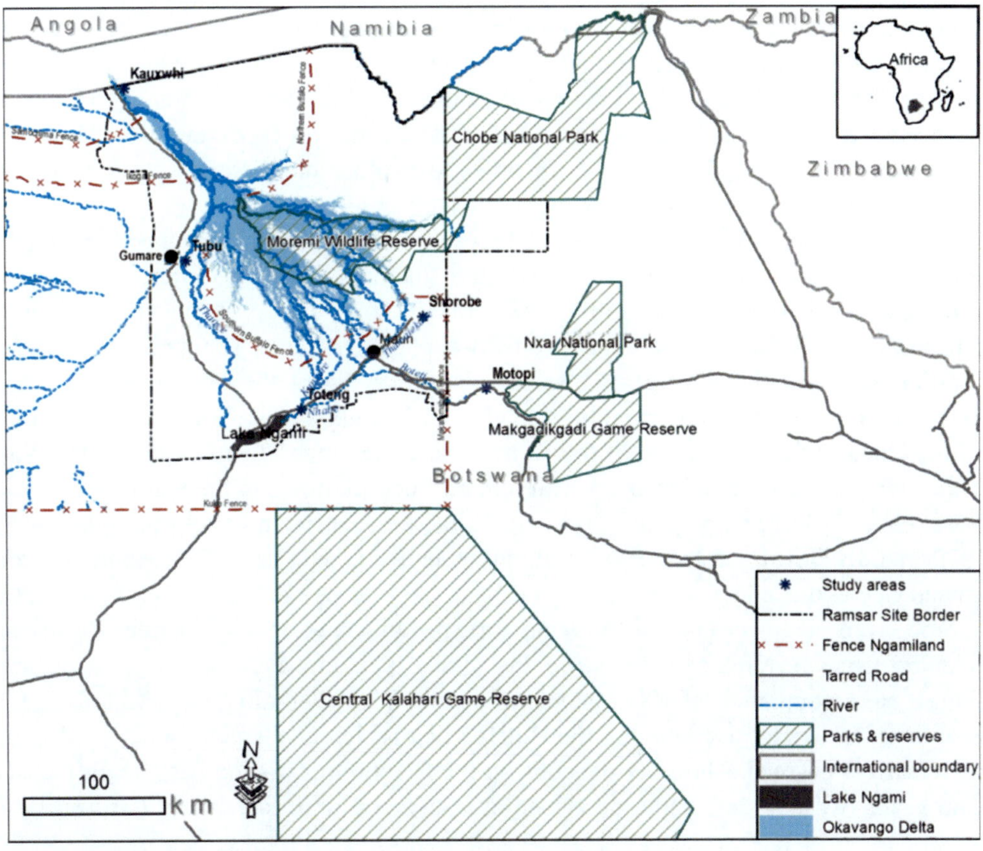

Figure 8.1.The Okavango Delta and the study villages.

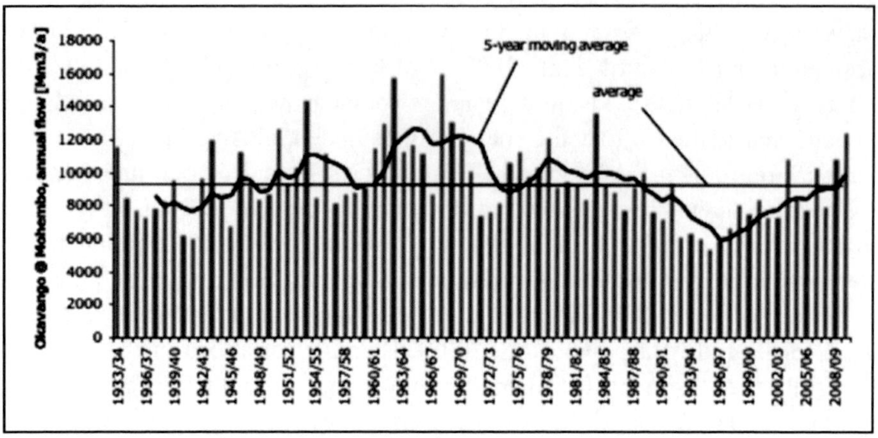

Source: Department of Water Affairs and HOORC (2010).

Figure 8.2. Variations in annual inflow into the Okavango Delta for the period 1933-2010.

The resultant flood may have attracted lesser public and research attention than the floods resulting from the 2004, 2009 and 2010 inflows. The inflows measuring between 11 000 Mm^3 and 12 000 Mm^3, and the resultant floods occurred at a time when the entire Delta had increased human activity with higher population concentrations, more economic activities and major land use changes. The Delta is a habitat for many aquatic and terrestrial species as well as an important livelihood resource base for the human population in the entire Ngamiland District (McCarthy *et al.*, 2003; Kgathi et al., 2004).

The specific study sites are the villages of Kauxwi, Tubu, Shorobe, Toteng, Motopi and their respective associated localities of arable lands and cattle posts (Figure 8-1). These were selected based on geographical location in order to explore the different conditions under which households adapt to extreme flooding in the Delta.

Kauxwi is located in the eastern side of the upper panhandle* within an area covered by permanent floodplains, often affected by extreme flooding. The village and its associated localities have 1 313 people of mainly Hambukushu ethnic group. Tubu village is in the western mid-Delta, within a wide expanse of flood plains and river channels. This area has been partially desiccated for many decades due to the drying of the Thaoge River. It has a total population of 754 people of mainly WaYei ethnic group. Toteng is in the lower western part of the Delta at the confluence of the Kunyere and the Nhabe rivers. Most of the Nhabe River channel can be described as deep with limited flood plain while the Kunyere is shallow and has a relatively large flood plain. These rivers had desiccation in various periods since the late 1970s, hardly flowing at the same time until recently. Toteng has 2 513 people of mainly Herero and Gcereku ethnic groups.

Shorobe is in the lower Okavango Delta, northeast of Maun. It is surrounded by seasonal floodplains that have remained dry since the 1980s and now being affected by extreme flooding. It has a population of 1 815 people of mainly WaYei ethnic group. Motopi is in the southern distal Okavango Delta along the Boteti River which is characterised by a deep and relatively narrow channel. The river depends on the Delta outflows and has been affected by desiccation for two decades, being described as ephemeral until late 2000s (Arntzen *et al.*,

* The narrow section where the Okavango River enters the Delta as commonly known.

1994; Mazvimavi and Motsholapheko, 2008). It has a population of 1 130 people who are mainly WaYei. Conditions for slow and wide coverage of the flood exist for both Tubu and Shorobe due to the extensive flood plain around these areas. Soil fertility in the Delta flood plains and river channels is higher than in the surrounding dryland, which is made up of the Kalahari sands. *Molapo* farming (flood recession cultivation) is mostly practiced in these areas and commonly so in the villages of Tubu and Shorobe than Kauxwi, Toteng and Motopi. All the study villages have on-site clinic, water reticulation system, primary school, and tribal administrative offices. Motopi, Toteng and Shorobe have tarred road links to major villages while others are linked by gravel road. The associated localities surrounding these settlements, generally have limited access to these infrastructural services.

Methods

This was a non-interventional analytical study, undertaken mainly by a cross-sectional survey of households, supplemented by desktop searches, interview of experts, key informant interviews and focus group discussions. The desktop searches involved the review of all relevant sources of secondary data. Some experts considered key to the study of livelihoods, floods and rural income, were also informally interviewed by email, telephone or face-to-face discussions. These served to verify data from some secondary sources and avoid technical pitfalls.

Sample Survey

The study used the household as a unit of analysis. The number and location of households in the villages were verified using enumeration area (EA) maps, updated for the 2011 national census. A list of all households in each of the study villages and associated localities was compiled. The total number of listed households was used to calculate the sample size (Table 8-1).

From a list of all households in the respective villages and associated localities, a simple random sample of 623 households representing 40% of the accessible population was drawn. In behavioural research, samples of not less than 5% of total or not less than 30 units are considered sufficient for statistical analysis (Moser and Kalton, 1971; Balnaves and Caputi, 2001). The methods of sampling as illustrated above were also used in a pilot study undertaken in December 2009. It included a pre-test of the questionnaire, key informant interviews and focus group discussions. These were conducted in Makalamabedi (Central), Sehitwa (near Motopi and Toteng, respectively) and Shorobe. The results have helped to improve the quality of the data collection instruments and sampling procedures that were used in the household survey.

The household survey was conducted from February to August 2010. A questionnaire was directly administered to the selected households. It comprised of structured and semi-structured questions on a) household demographic and socio-economic background, b) impacts of extreme flooding on households, and c) household coping and adaptive strategies.

Table 8.1. Population size, number of households and sample size for each of the study villages and their associated localities

Village	Population size*	Total number of households listed	Number of households sampled
Kauxwi	1313	243	97
Tubu	754	188	75
Toteng	2513	584	234
Shorobe	1815	356	142
Motopi	1130	188	75
Total	7525	1559	623

Source: *Central Statistics Office (2002).

Measuring Access to Capital

Household access to various forms of capital was measured using a number of proxy indicators with different units such as minutes (time), cattle equivalent units (livestock) and percentages derived from household responses relating to ownership or access to capital. Similar principles applied by Hahn *et al.* (2009) in the development of the livelihood vulnerability index were used with modification to focus on household capacity to adapt. A set of unit-free indices of values between zero (lowest) and one (highest) were developed in order to compare household access to various forms of capital. These indices were calculated using equation (1) adapted from the standardisation of all indicators comprising the human development index (HDI) developed by UNDP (1990):

$$Index_a = \frac{\alpha - a_{min}}{\alpha_{max} - a_{min}} \tag{1}$$

where a is the value of the indicator measured in any unit, a_{min} and a_{max} are its pre-determined minimum and maximum values. The allocation of minimum and maximum values was done as follows: a) values measured in percentage ranged from zero to 100, b) the average borrow-lend-money ratio measuring access to bonding social capital varied from 0.5 to two (household that borrowed but never lent scored 2:1, and vice versa), c) time indicating access to health ranged from three to 300 minutes, and d) ownership of livestock measured in cattle-equivalent-units (CEU)[†] adopted values of the lowest and highest number of cattle-equivalent livestock owned in all the study villages (zero and 529) respectively. For clarity two things here deserve mention, 1) livestock was included as one of the indicators of financial capital since it was identified as a source of cash, savings or insurance when sold or exchanged by households to meet immediate requirements and 2) the measurement of social capital included the three components of bonding, bridging and linking social capital. Bonding social capital are the reciprocal exchanges or assistance emanating from close ties between relatives and friends. Bridging social capital are those exchanges derived from membership to local organisations, while linking social capital are exchanges obtained through links beyond the community (Stone, 2001; Woolcock, 2001).

[†] One cattle-equivalent-unit is equal to six goats/sheep or 0.8 horses/donkeys in Botswana.

All the indices were averaged for each type of capital using equation (2) below:

$$Capital_a = \sum_{i=1}^{5} \frac{Index_{ai}}{n} \quad (2)$$

where *capital*$_a$ is type of capital, *index*$_{ai}$ is the index of any of the indicators calculated using equation (1) and n is the total number of indicators used to estimate *capital*$_a$. Households were also asked to state their livelihood activities and rank them according to four levels of "most important", "second-", "third-", and "fourth important". Food security was measured in terms of the number of months in a year that households found it difficult to obtain food.

RESULTS AND ANALYSIS

The overall sample comprised of 55% male-headed, 33% *de jure* female-headed[‡] and 12% *de facto* female-headed households. The households had a mean of 7.1 people in an extended family structure. Households with no formal education made up 56%, which were mainly male-headed (32%), while *de jure* and *de facto* female-headed households comprised 19% and 4%, respectively.

Household Access to Capital

The general distribution of the responses indicates that households had high access to natural capital. The proportion of households which stated that there were no factors that impeded their access to land for livestock rearing and other uses ranged from 71.6% in Toteng to 87% in Shorobe. Ownership of arable land ranged from 70% in Toteng and 93.2 % in Tubu. The proportion of households who said they had unimpeded access to water for productive purposes ranged from 89% in Toteng to 95% in Tubu, and those who stated that they have unimpeded access to forest resources ranged from 71.6% in Kauxwi to 86.2% in Shorobe (Table 8-2).

Access to financial capital was generally low in all the villages except in Shorobe and Motopi where more than 50% of households had access to credit and insurance. Households in these villages accessed these through self-help borrowing schemes (*motshelo*) and funeral benefit schemes, respectively. Access to cash was the lowest, ranging from 11.5% of households in Toteng to 48.9% in Shorobe. The mean value of cattle ranged from 4.3 to 37.1 cattle-equivalent-units in Kauxwi and Toteng respectively. Households that indicated that they had difficulty accessing financial capital also stated that they found it hard to pay school fees, buy school uniform for their children as well as acquire human necessities such as food and clothing. The following were identified, in focus group discussions, as factors contributing to low access to financial capital: 1) foot-and-mouth disease control regulations, 2) lack of markets for agricultural products, 3) lack of credit and insurance markets, and 4) general poverty.

[‡] *De jure* female-headed refers to households directly headed by females including widowed, never married and so on. *De facto* female-headed are those headed by females while the spouse or partner is indefinitely or permanently away.

Table 8.2. Determinants and indicators of access to capital in all the study villages

Type of capital and indicators	Unit	Kauxwi	Tubu	Shorobe	Toteng	Motopi
Natural capital						
Land: households stating unimpeded access	percent	83.3	73	87	71.6	70.7
Water (productive purposes): households stating unimpeded access	percent	89.5	95.8	94.9	89	88
Forest resources: households stating unimpeded access	percent	71.6	75.7	86.2	81	68
Household ownership of arable field	percent	90.6	93.2	85.8	69.8	81.3
Financial capital						
Cash: households stating unimpeded access	percent	36.5	31.6	48.9	11.5	44
Credit: households stating unimpeded access	percent	39.4	30.6	53.2	43.2	72
Insurance: households stating unimpeded access	percent	49.5	37.7	61.6	41.5	69.3
Livestock: mean number in cattle equivalent units (CEU)	CEU	4.3	30.8	16.3	37.1	30.8
Social capital						
Bonding capital: a) labour exchanges	percent	42.3	58.7	71	71.4	38.7
b) average borrow-lend money ratio (range 0.5-2)	ratio	1.02	1	1.02	0.99	0.97
Bridging capital: membership to local organisations	percent	25.8	20	19.1	25.2	36
Linking capital: a) households seeking assistance from village extension	percent	63.2	62.7	63.6	64.8	69.3
b) households seeking assistance from local/district authorities	percent	28.9	24	29.6	39.3	25.3
Human capital						
Proportion male-headed	percent	61.9	70.6	59.9	69.8	62.7
Head with formal education	percent	51.6	33.3	36.4	48.4	37.3
Households with non-agricultural skills	percent	54.9	53	65.4	39.2	50.7

Type of capital and indicators	Unit	Kauxwi	Tubu	Shorobe	Toteng	Motopi
Households with member working elsewhere	percent	38.4	36	52.8	39.9	42.5
Households reporting good health	percent	37.2	46.7	47.9	51.9	44
Physical capital						
Health facilities: average time to health facility	1/n minutes	0.022	0.014	0.012	0.018	0.063
Water reticulation: access to piped drinking water	percent	79.4	52	47.9	48.7	85.4
Tools of the trade: plough, donkey cart, crop storage	percent	40.2	58.2	41.3	28.1	37.1
Communication: radio, cell phone	percent	43.3	62.7	74	62.8	64.7

Table 8-2 also shows that access to bonding social capital in the form of manual labour exchanges was low in the two villages of Motopi (38.7%) and Kauxwi (42.3%) but high in the other villages. Few households had borrow-lend-money arrangements with their relatives and friends. The average borrow-lend-money ratio in all the study villages was close to one in all the villages. According to the World Bank (1997), this ratio varies between the lowest value of 0.5, when a household borrowed less than it lent money, and the highest value of two when the household borrowed more than it lent money.

Access to bridging social capital was also very low in all the villages as shown by the low proportion of households with membership to local organisations (Table 8-2). Access to linking social capital was high for extension services but low for local and district authorities, which included traditional leadership, local councillors and district administration (Table 8-2). Focus group discussions conducted in the lower and distal Delta have also revealed that bonding social capital has generally declined, being limited to cultural ceremonies such as funerals and weddings. Other indigenous practices that supported bridging and linking social capital such as collective labour arrangements (in *letsema or lesotla*), livestock loaning (*mafisa*) and age regiments are no longer practised. According to the discussants, this may be attributed to the gradual loss of culture, growing mistrust among kin and friends, and the general desire to work as independent individuals.

Access to human capital was generally low as shown by the low proportions of households with formal education, good health and those with members engaged in non-agricultural activities outside their village. Most households were male-headed but did not have formal education except in Kauxwi (51.6%). Good health was stated in less than 50% of households in all the villages except Toteng (51.9%). The proportion of households with a member engaged in non-agricultural activities in other settlements was lower than 50% in all the villages except Shorobe. This village is located closer to the district capital Maun where opportunities for employment in non-agricultural activities were better than in most rural areas.

Household access to physical capital was also low. Average time taken by households to reach health facilities was high in all the villages except Motopi. It ranged from 45 to 83 minutes compared to 15 minutes in Motopi. This village did not have many outlying localities and household residences centrally located. The proportion of households that had piped drinking water was high in Motopi (85.4%) and Kauxwi (79.4%) but low in the other villages. Access to telecommunications, such as the radio and cellular phones was high in all the villages except in Kauxwi (43.3%). It ranged from 62.7% of households in Tubu to 74% in Shorobe. Tools of the trade such as ploughing equipment and grain storage were accessible to less than 50% of households in all the villages except in Tubu. Informal interviews revealed that those without equipment relied on borrowed or hired equipment and subsequently delayed ploughing.

These indicators were converted to indices and compared by study village. Hypothetically, the community with the highest access to all forms of capital will have an index of one and the lowest equals to zero. This implies the respective widening and narrowing of the capital access pentagon as summarised in Figure 8-3.

Access to natural capital was high for all the study villages, the highest being in Shorobe with an index value of 0.89. The indices for access to other forms of capital were generally low with slight variations by villages. Access to human capital was almost the same for all the villages though Shorobe showed a slight edge over the other villages. Toteng had the lowest access to natural (0.78) and physical (0.39) capital compared to the other villages. In case of social capital, Toteng (0.47) had a slightly higher access value than the other villages while Motopi (0.35) had the lowest access to this capital. Access to financial capital was slightly higher in Motopi (0.48) than in the other villages, while Toteng (0.26) and Tubu (0.26) had the lowest values.

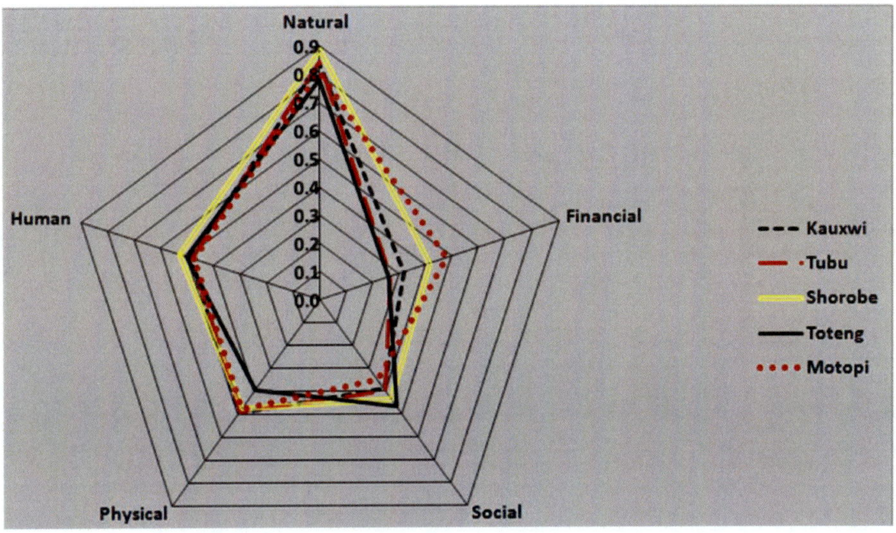

Figure 8.3. Variations in access to capital in the study villages.

Household Livelihood Activities

Households engaged in an average of three livelihood activities, a maximum number of nine activities and very few households (8%) had one livelihood activity. Poor households had fewer livelihood activities, mostly supplemented by social welfare and other government assistance schemes. Households combine subsistence arable farming (*molapo* or dryland farming) and livestock with other livelihood activities such as formal employment, informal employment, drought relief employment/labour-intensive-public-works (LIPW). The distribution of livelihood activities by study village (Figure 8-4) indicates that rural households have major differences in the way they combine livelihood activities.

These differences may be related to household cultural background and the types of resources available in the respective locations. Livestock farming was the most common, being practiced by more than 50% of households in four of the five study villages. However, livestock was not commonly practiced in Kauxwi in the Delta Panhandle, where most of the households (81%) practiced dryland arable farming. It is also surpassed by *molapo* farming in Tubu, in the mid-Delta. The villages of Kauxwi and Tubu are dominated by the Hambukushu and WaYei ethnic groups that mainly practice dryland and *molapo* arable farming, respectively. Crops commonly planted in these villages include maize, beans, sweet reed and melons, though sorghum and millet are rarely planted by *molapo* farming households. When livelihood activities were ranked in terms of levels of importance, livestock farming was stated as the most important in the villages of Tubu (41%), Toteng (50%) and Motopi (38%).

Regardless of the higher proportion of households that practice *molapo* farming compared to livestock, Tubu households maintained that livestock farming was the most important. This may be related to the high risks associated with *molapo* farming compared to livestock. *Molapo* farming is prone to crop damage by annual floods, livestock and wildlife. Dryland farming was ranked as the most important in Kauxwi (60%) though it is prone to crop failure due to unreliable rainfall and poor soil fertility. Most of the Delta is covered by infertile sandy soils with low moisture retention capacity.

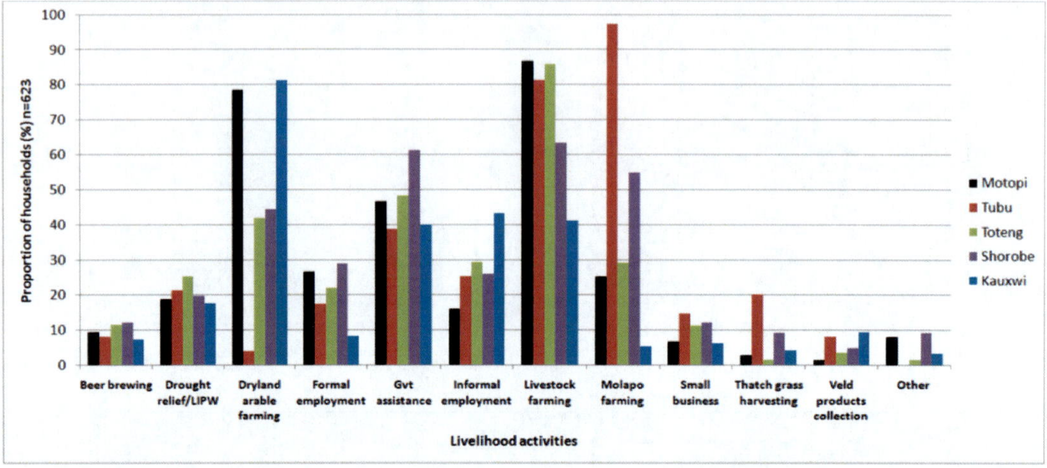

Figure 8.4. Proportion of households by types of livelihood activities in the study villages.

Government assistance surpassed most non-agricultural activities in all the villages except in Kauxwi. It was the most dominant after livestock, particularly in the villages of Toteng and Shorobe in the lower Okavango Delta. In effect, households in Shorobe indicated that it was the most important livelihood activity. Fishing was practiced by a small proportion of households (0.4%) in all the study villages, contrary to expectation that it may be one of the most practiced livelihood activities, in the mid- and upper-Delta. Fishing and veld products collection were not mentioned as most important in all the study villages. Other livelihood activities undertaken include backyard gardening, hair cutting/braiding, poling and basket making, all of which served as household safety nets.

Impacts of Extreme Flooding on Household Livelihoods

Household survey results indicate that in 2004, seven percent of all households interviewed were affected by extreme flooding. These were in the villages of Kauxwi and Tubu in the upper- and mid-Delta, respectively. Some households lost their crops, while others were displaced or lost property and livestock. More female-headed households (60%) were affected by this flood than male-headed households (40%). The Pearson's chi-square of independence showed an association between gender and the affected households in 2004. It had a value of 4.71 ($p=0.03$, at one degree of freedom,) significant at five percent level.

Disruptions were also observed in 2009 with 23% of households being affected and 61% of them being male-headed. The main impacts of extreme flooding were crop damage (53%), household displacement (18%), destruction of household property (12%), livestock drowning and mud-trapping (11%), and others (6%) including loss of human life due to drowning (Figure 8-5).

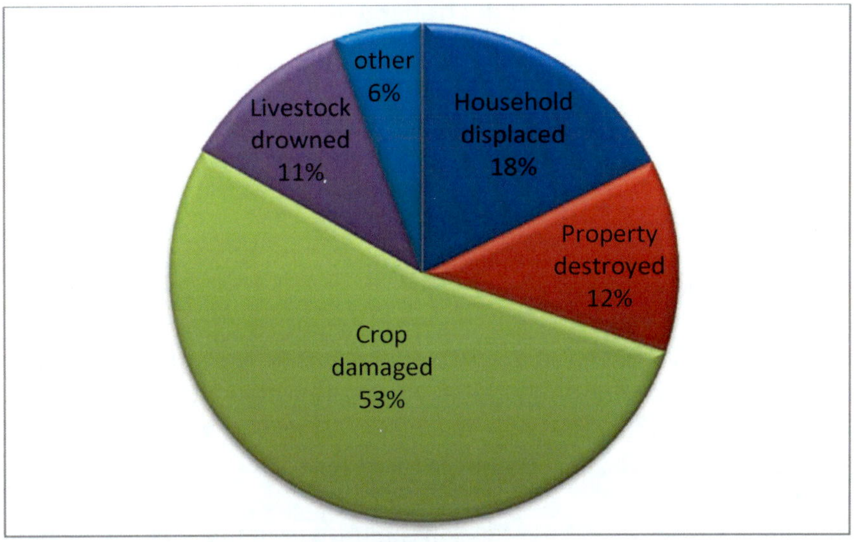

Figure 8.5. Proportion of households variously affected by extreme flooding in 2009 in all the study villages.

Most crop damage occurred in the villages of Tubu and Shorobe affecting 27% and 47% of all households who lost crops, respectively. In the other villages this proportion varied from three percent in Motopi to 14% in Toteng.

Data from the District Administration office indicate that in 2009, extreme floods displaced 157 households (856 people) in twelve villages in the upper Delta, including the study village of Kauxwi (District Administration Ngami sub-District, 2010). The displacement of households occurred throughout the Delta and included areas that were never affected by floods in many years. However, it affected more households in communities in the upper Delta than other areas. Generally, Kauxwi and its neighbouring villages of Xakao and Mohembo have been affected by floods annually with minimal displacements.

However, in 2009 and 2010 these villages had more household displacements, crop damage and loss of property with impacts on livelihood activities. A total 25 households (118 people) were displaced in Kauxwi, 17 (169 people) in Xakao and 14 (59 people) in Mohembo. In both Kauxwi and Xakao, the displacement resulted in permanent relocations to less risky areas as some affected households moved to newly allocated residential plots (District Administration Ngami sub-District, 2010). For the first time in many years, a total of 31 households (150 people) were also displaced in Toteng and associated localities. An estimated 109 hectares of *molapo* fields, for 67 households, were submerged by the floods in five villages of the upper Delta including Kauxwi (District Administration Ngami sub-District, 2010). As later confirmed by key informants, this resulted in food shortages for the affected households.

In localities around Toteng, some households stated that flooding reduced livestock grazing pastures as most of the flood plains were submerged for many months. This resulted in livestock deaths due to starvation and mud-trapping. Some households that were evacuated to enhance access to basic services left large numbers of livestock in islands such as Morula, Matsebe 1 and Matsebe 2 (Table 8-3).

Table 8.3. Number and type of livestock abandoned in some affected settlements

Settlement	Livestock type and number affected		
	Cattle	Goats	Sheep
Morula	1237	274	58
Matsebe 1	1050	170	25
Matsebe 2	1016	234	76
Total	3303	678	159

Source: DA Ngami sub-District (2010).

This made it difficult to monitor the status of such abandoned livestock with possible rise in predation, mud-trapping and straying. In some cases flooding separated households, weakening social ties and mutual assistance among kin. This mostly disadvantaged vulnerable groups such as the elderly, some of whom abandoned their livestock to live with relatives on the other side of the river.

The destruction of crops in *molapo* fields has mostly been noted in the study villages of Shorobe, Tubu and their respective associated localities. It has been described by households as severe in some cases. Household interviews indicate that 72% and 67% of households in Shorobe and Tubu, respectively, had whole field crops destroyed by flooding. Most of these remained submerged for prolonged periods of time reducing the ploughing period with adverse effects on household food security. The impact of other shocks on *molapo* farming such as crop damage by elephants have also been identified as significant in the lower Delta. Some households indicated that most of their crops were destroyed before the flood reached them. As also confirmed in key informant interviews, the losses adversely affected household food security and the ability of households to deal with the impacts of extreme flooding.

At the community level, extreme flooding disrupted some services throughout the Delta with effects on households in the study villages. For instance, the floods disrupted three water treatment plants, causing water shortages in 22 villages of the Okavango sub-District including Kauxwi and Tubu. Some households in Tubu resorted to using river water for domestic purposes including drinking. In the lower parts of the Delta, some boreholes were also shut down after being submerged, resulting in widespread water shortages and tensions among residents in some settlements. In Somelo, a locality of Toteng, local residents accused public officers of taking more than a fair share of the water supply by filling large water tanks. The settlement was supplied through water bowsers that came once in two days.

Road transport services to major settlements were also disrupted after the main bridge in Toteng was submerged and remained so for two months. This prompted district authorities to reconstruct old gravel roads to divert traffic. Road links to some localities in Toteng, Tubu and Shorobe were also cut out urging some residents to relocate in order to access health, transport, education for children and most other basic services. In some localities such as Ditshiping and Morutsa in Shorobe, most community members especially the elderly stayed for some months without health and social welfare services. Community tourism activities were also disrupted in these areas with subsequent loss in household income. Informal interviews indicated that a total of 240 community-based *mokoro* (canoe) polers and tourist guides spent three months without tourist clients. This resulted in loss of opportunities to earn monthly income, more so as it occurred during the tourism peak season (March to September). During that time, a poler could earn up to BWP1 500 (approximately US$200) a month. Households in these localities depended on daily cash earnings from poling tourists on *mokoro* excursion trips.

Household Coping and Adaptive Strategies

Rural households in the study villages used different coping strategies depending on the impacts and the affected livelihood activities. The following main coping strategies were used by varying proportions of households in response to extreme flooding in 2009, a) labour switching to other livelihood activities b) seeking government assistance and c) short-term local mobility (Figure 8-6).

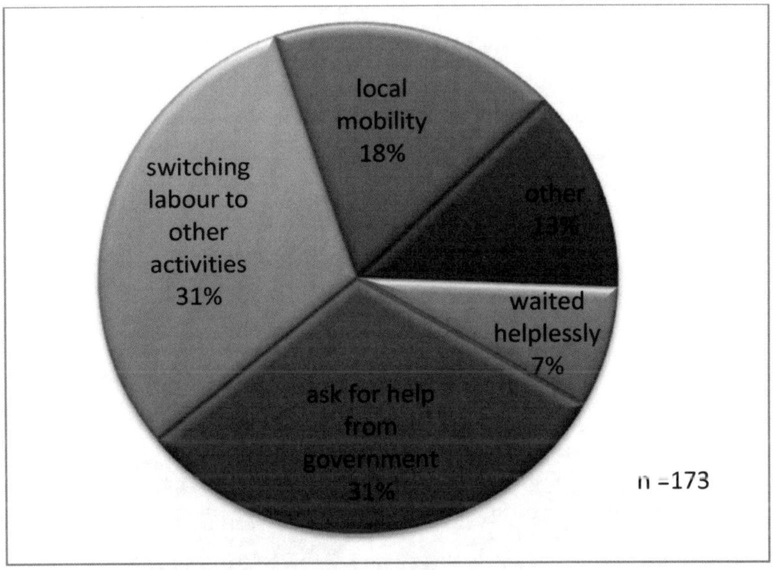

Figure 8.6. Household coping strategies in the study villages.

Labour switching was used by 31% of *molapo* farming households and this was mostly in Shorobe and Tubu. Government assistance was used by 31% of households most of which were displaced by the floods in Kauxwi and Toteng. The assistance was in the form of evacuations, temporary and permanent shelter, provisional food rations and clothing. Permanent shelter and clothing were provided to the most vulnerable. Seven percent of households stated that they "waited helplessly" for several days before being assisted by the Government. Some of these were variously assisted by relatives and friends, while others stayed without food over the period. These were mainly the elderly in female-headed households, and those who were delisted from the destitute food ration scheme. Short-term local mobility was used by 18% of households, mainly in Toteng and Kauxwi. Some households temporarily moved to stay with relatives, or to reside at own place in other locations, re-deploying their labour in other livelihood activities. Others relocated their cattle-posts to less affected areas.

Other coping strategies included the use of canoes to harvest crops in submerged *molapo* fields and to rescue livestock mainly goats trapped in islands. This was undertaken by 13% of households which were mostly male-headed. These were in the study villages of Tubu, Shorobe and Kauxwi. The former two villages comprised of the WaYei and the latter was mainly inhabited by the Mbukushu (of the Panhandle) ethnic groups. These two ethnic groups are well renowned for their water navigation skills.

Long-term adaptive strategies identified in the study villages were livelihood diversification and long-term local mobility done in anticipation of the shock or in response to the impacts, respectively. As shown in section 3.2 most households had diversified livelihood portfolios. The multiple livelihood activities made it possible for them to cope through labour switching.

Households did not state migration as their adaptive strategy against extreme flooding. They indicated that flooding did not cause any major movement but resulted in temporary relocations within the same area. However, migration was stated as a strategy against other shocks such as desiccation of river channels. The results also showed that 41% of all households in the study villages had a member who went to other settlements for employment. Though this could not be directly linked to extreme flooding, it may serve as a useful adaptive strategy in case of any livelihood shock. Households also invested in non-agricultural skills that could be used to earn a living in case of on-farm livelihood decline or failure due to shocks, such as extreme flooding. These included the crafts/artisanal, professional (education and other sectors), elementary, service/clerical/sales, plant/machine operating, fishing and other skills such as hair styling, fashion design, music and performance arts (Figure 8-7).

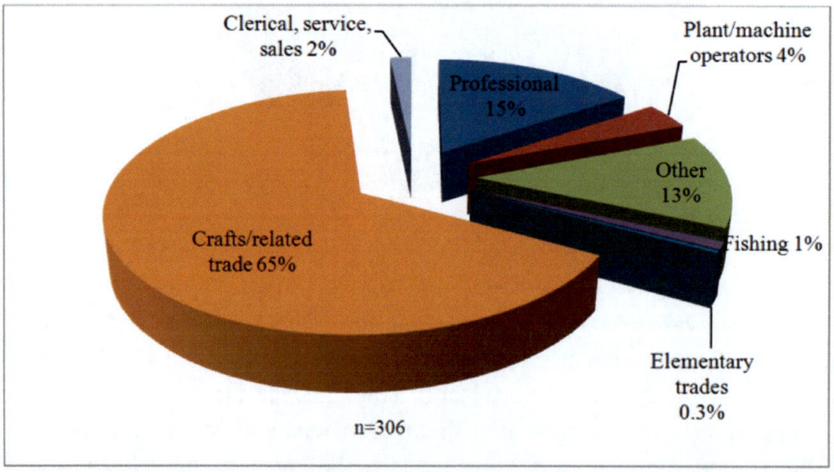

Figure 8.7. Types of non-agricultural skills possessed by households in the study villages.

With regard to pre-shock risk avoidance, 47% of the households, which were affected by extreme flooding, indicated that they received early warning about the flood in 2009. The District Disaster Committee had sent early warning through public announcements in the radio and at *kgotl*[1] meetings, poster displays in public places and short messages to cellular phone users. Some households stated that they heard about the upcoming flood from other community members. Among households that received early warning in 2009, 56% stated that they did nothing to prepare for the flood. Most of these were in Toteng (28%), followed by Shorobe (11%), Tubu (8%), Kauxwi (6%) and Motopi (3%). Some indicated that they were not sure of the timing between early warning and the arrival of the flood or did not believe that the flood will reach their areas. Actions undertaken by those that acted on early warning included early harvesting of crops, livestock relocation, early evacuation of some household members and others.

DISCUSSION

Household access to natural capital, such as land, surface water and some forest resources, was high in all the study villages. This partly explains the high dependence on natural resource-based livelihoods, as shown by the results, in the study villages. It further confirms the findings and observations made in other studies, that rural communities depend on the Delta's rich natural resources (Bendsen and Meyer, 2003; Kgathi *et al.*, 2004; Mbaiwa, 2004; Mendelsohn and el Obeid, 2004; Magole and Thapelo, 2005; Vanderpost, 2006; Wilk and Kgathi, 2007). However, access to some forest resources, such as thatch grass and some edible plants was relatively low in all the study villages. This also confirmed observations that these were scarce around settlements in the Okavango Delta (Applied Development Research Consultants (ADRC), 2001; Kgathi *et al.*, 2005).

Access to land, surface water and forest resources is subject to the communal land tenure system. This allows community members free access to grazing pastures, arable land and forest resources for various natural resource-based livelihood activities. However, there is mounting evidence suggesting limited access in future due to land use and policy changes, animal disease control fencing, population growth, upstream developments and global economic changes (Vanderpost, 2006; Wilk and Kgathi, 2007; Magole, 2009; Magole and Magole, 2009). Currently, the impacts may not be felt by households due to counteracting processes, such as deagrarianisation and population redistribution. Deagrarianisation, the process of occupational shift away from agrarian livelihoods, has reduced farm-based activities, and resulted in population redistribution from rural to urban settlements in many countries of sub-Saharan Africa, including Botswana (Ellis, 2000; Bryceson, 2002; Kgathi *et al.*, 2004; Vanderpost, 2006).

Households had limited access to financial, physical, human and social capital. The low access to financial capital may be related to: 1) the subsistence nature of livelihoods in the study villages, 2) limited markets for agricultural products, 3) general absence of credit and insurance markets, and 4) the frequent outbreak of foot and mouth disease. The latter may have contributed more to the low access to financial capital in Toteng and Tubu. These villages, and the surrounding localities, were in the core area of the outbreak and the sale or movement of all cloven-hoofed livestock and related products were banned. In the other villages localised movement and sale was allowed, while the village of Motopi was not affected these regulations as it falls within a separate disease control zone. This low access

[1] The *kgotla* is the traditional place and forum for discussing community issues in Botswana.

made it difficult for some households to meet their basic needs or improve current livelihood activities to deal with future incidences of extreme flooding.

Physical capital in the form of roads, health facilities, water reticulation systems and telecommunications, was accessible to most households in the main settlements. However, access was low in the associated localities with impacts on social wellbeing and capacity to deal with extreme flooding. Households had low access to tools of the trade and therefore low capacity to derive maximum benefits from the main livelihood activities. Access to human capital was also low as shown by low education, poor health and the general lack of marketable skills.

The generally low access to financial, physical and human capital in the study villages is typical of most rural areas in the developing countries of Africa and Southeast Asia (Ellis, 2000). However, the low access to social capital as shown in this study is atypical of these trends. This low access may be due to disintegration of cultural norms, breakdown in household social fabric, economic transformation and the general collapse of some indigenous institutions that promoted bonding among households. Dube and Sekhwela (2008) also identified some of these as the underlying factors leading to reduced household resilience to shocks in some parts of eastern Botswana. Kgathi *et al.* (2004) also found that the powers of traditional leadership and influence of the *kgotla* system have been generally degraded by the creation of modern institutions in the Okavango Delta.

Extreme flooding mainly affected households through crop damage in *molapo* fields, livestock drowning and mud-trapping, household displacement and damage to property. *Molapo* farming was not considered by households as the most important livelihood activity though it was a source of food. This may be due to the high risk of loss associated with annual flooding, crop damage by livestock and wildlife, and insecurity of tenure. Livestock drowning and mud-trapping mostly affected households in Toteng and associated localities. This may have been due to loss of memory about the flooding patterns of the area, or insufficient knowledge about the flooding cycles of the Delta. Some of the affected households in this area were among those that moved in during the prolonged desiccation of Lake Ngami. This occurred in the mid-1990s and early 2000s, compelling households to relocate inward into the Delta in search of water for livestock (Kgathi *et al.,* 2004).

Early warning did not reach all households in the study villages and even among those that received it, there were uncertainties about timing or the spatial extent of the flood. Lack of knowledge about flooding patterns of the Delta was not limited to households alone. Even in the public sphere, it may have largely contributed to the impacts of flooding on public infrastructure and other services. Most infrastructure developments were undertaken in the 1980s and 1990s, during the dry phase of the Delta and were suitable for the conditions thereof. For example, major public roads had small culverts constructed in main river channels, such as the Kunyere near the study village of Toteng. Some schools were constructed in multiple-island villages like Tubu without any culverts linking them to the main village. Boreholes providing water to some villages were sunk in flood plains and were submerged during flooding in 2009, 2010 and 2011.

Households coped with the impacts of extreme flooding through labour switching to other livelihood activities, temporary relocation to less affected areas and use of canoes for either early harvesting or for recuing their livestock from drowning. The switching of labour to other activities during or after the impact of extreme flooding indicates that households wisely allocate labour to other risk spreading activities in order to bounce back after the shock. This coping mechanism is the most common as households have numerous livelihood activities that sustain them. Temporary relocation to less affected areas indicates the importance of maintaining high access to natural capital as this cannot be possible without more access to land, grazing pastures and woodland resources. The use of canoes was limited to those households and communities with water navigation skills related to culture. It was also used by households in relatively wet areas, while those in dry areas seemed to have lost the skill due to non-use in the many decades of desiccation. The choice of anticipative actions such as early harvesting of crops and relocation of some livestock may have been influenced by early warning.

The main adaptive strategies were livelihood diversification, long-term local mobility, training in non-agricultural skills and internal migration within and across districts. The investment in non-agricultural skills that occurred before the extreme floods enabled some households to diversify their livelihoods and withstand the impacts of extreme flooding. Studies in the Okavango Delta have shown that rural households engage in migration for paid employment and local mobility to less affected areas in response to river desiccation and drought (Vanderpost, 1991; Bendsen and Meyer, 2003; Kgathi *et al.,* 2004; Wilk and Kgathi, 2007).

These adaptive strategies have been used to provide livelihood security and coping with multiple shocks in the Okavango Delta and Botswana in general (Silitshena, 1982; Tlou, 1985; Scudder *et al.*, 1993; Bendsen and Meyer, 2003). Historically, migration to South African mines and locally, provided income to rural households for general household needs as well as investment in agriculture and acquirement of other forms of capital (Valentine, 1993; Reardon, 1997). Today, migration is still useful as it provides income to rural households, access to other forms of capital such as education for children (human capital), purchase of physical assets, and opportunities to recover from the impacts of shocks including extreme flooding (Kgathi *et al.*, 2007).

Livelihood diversification, as an adaptive strategy, is evident in livelihood studies undertaken in the Okavango Delta (Scudder, 1993; Fidzani *et al.,* 1999; Cassidy, 2003; Kgathi *et al.*, 2007; Swatuk and Motsholapheko, 2008). This helps households to, avert risk and cope with the impacts of multiple shocks including extreme flooding (Wilk and Kgathi, 2007).

CONCLUSION

The impacts of extreme flooding on household livelihoods may be significant despite the generally positive view of flooding in the Okavango Delta. The adverse impacts of extreme flooding on *molapo* farming, livestock keeping and the disruption of service infrastructure provide good lessons for adaptation to climate variability and change in the Okavango Delta and beyond. For the Okavango Delta, extreme flooding cannot be a major threat to household livelihoods if proper planning and prevention of memory loss are maintained through knowledge transfer in both the public and private spheres. In the public sphere, knowledge transfer can be promoted through building of databases to support planning and policy, while at the household level this can be inscribed in oral traditions. This study adds to the former and the methods used in the estimation of household access to capital and adaptive capacity can be copied or modified to develop better and universally applicable tools.

In the Okavango Delta, and particularly in the study villages, rural household capacity to adapt largely depends on access to natural capital. Households maintained multiple and mostly natural resource-based livelihood activities and this helped them to survive the collapse or disruptions in other activities. However, this is threatened by the growth in population, land use changes, policy shifts, upstream developments, global economic changes and flood variations due to climate variability and change. A major policy challenge lies in maintaining high access to natural capital in the face of population growth, its related land use changes and the need for resource conservation in the Okavango Delta as a Ramsar site.

The study identified household coping strategies, such as labour switching and local mobility, and adaptive strategies such as livelihood diversification and migration, as the mainstay of household adaptation in the study villages. This information may provide an entry point for planned interventions to address the impacts of other shocks now and in the future. Studies have shown that in most cases, the solutions of impacts of one shock may also be applicable to other shocks. In the Okavango Delta, these strategies can be further developed to improve the adaptive capacity of vulnerable households, estimated at 38% in this study. This will not only be a solution for addressing the impacts of flooding shocks, but may also apply to other shocks such as the current recession or the HIV/AIDS pandemic. Since shocks occur when they are least expected over long time intervals, responses can be inbuilt within long-term planned adaptations to maintain resilience in household livelihoods and avoid loss of memory relating to past incidences of the shock.

What then are the wider policy implications of this study on adaptation to shocks? To start with, the study may assist in providing information necessary for the design of policy interventions for the impacts of flood-related shocks in the Delta and other regions of similar nature. When supplemented by early warning systems, policy makers can be in a position to design effective strategies that can prepare communities to address the risks of extreme floods. The information may also provide a robust framework for addressing the future changes of climate in the Okavango Delta and other regions. Projections about future climate are still uncertain, but most climate change models predict that arid and semi-arid areas such as the Okavango Delta are likely to become drier in the future suggesting that the impacts of floods may become less severe (IPCC, 2008). However, the results of this study provide useful lessons to policy makers in areas likely to experience extreme flooding as result of climate change. Although the current impacts do not necessarily reflect those of climate change, they provide useful lessons about the impacts likely to be associated with climate change in the future.

Acknowledgments

This work is part of a proposal submitted to the Okavango Research Institute, University of Botswana for the degree, PhD Natural Resources Management. Funded by Carnegie-RISE and University of Botswana Office of Research and Development. Special thanks to: Dr. O.T. Thakadu, Professor B. Ngwenya and Professor P. Wolski for some insights. Further thanks to

J. Olefile and G. Mpiping for field data collection, the study communities, and all those who contributed in various ways.

REFERENCES

Anderson, L., Gumbricht, T., Hughes, D., Kniveton, D., Ringrose, S., Savenije, H., Todd, M., Wilk, J., Wolski, P., 2003. Water flow dynamics in the Okavango River Basin and Delta: a pre-requisite for the ecosystems of the Delta. *Physics and Chemistry of the Earth,* 28, 1165-1172.

Arntzen, J., 1989. Environmental pressure and adaptation in rural Botswana. *Unpublished PhD thesis.* Vrije University of Amsterdam, Amsterdam.

Arntzen, J., Chanda, R., Nkambwe, M., Ringrose, S., Sefe, F.T.K., Vanderpost, C., (1994). Desertification and possible solutions in the mid-Boteti River area: a Botswana case study for the intergovernmental convention to combat desertification (INCD). *Ministry of Agriculture,* Gaborone.

Ashley, C., Carney, D., 1999. *Sustainable livelihoods: lessons from early experience.* DFID, London.

Balnaves, M., Caputi, P. 2001. *Introduction to quantitative research methods: an investigative approach.* Sage Publications, London.

Bendsen, H., Meyer, T., 2003. The dynamics of the land use systems in Ngamiland, Botswana: changing livelihood options and strategies. *Paper presented at the Environmental monitoring of tropical and sub-tropical wetlands,* Maun, Botswana.

Berkes, F., Folke, C. 1998. Linking social and ecological systems for resilience and sustainability. In F. Berkes, Folke, C., Colding, J. (Ed.), *Linking social and ecological systems: management practices and social mechanisms for building resilience.* Cambridge University Press, Cambridge.

Brock, K., 1999. *Implementing a sustainable livelihoods framework for policy-directed research: reflections from practice in Mali.* IDS working paper no 90: Institute of Development Studies.

Cassidy, L., 2003. Anthropogenic burning in the Okavango panhandle of Botswana: livelihoods and spatial dimensions. *MSc Thesis,* University of Florida.

Central Statistics Office, 2002. *National population and housing census 2001.* Ministry of Finance and Development Planning, Gaborone.

Department of Environmental Affairs. 2008. *Okavango Delta management plan.* Department of Environmental Affairs, Gaborone.

District Administration Ngami sub-District, 2010. *Flood disaster report:* Nxaraga.

Dube, O. P., Sekhwela, M.B.M., 2008. Indigenous knowledge, institutions and practices for coping with variable climate in the Limpopo Basin in Botswana. In N. Leary, Adejuwon, J., Barros, V., Burton, I., Kulkarni, J., Lasco, R. (Ed.), *Climate and adaptation.* Earthscan, London, pp. 71-89.

Ellis, F., 2000. *Rural livelihood diversity in developing countries.* Oxford University Press, Oxford.

Ellis, F., Biggs, S., 2001. Evolving themes in rural development. *Development Policy Review,* 19(4), 437-448.

Ellis, F., Freeman, H.A., 2005. Conceptual framework and overview of themes. In F. Ellis, Freeman, H.A., (Ed.), *Rural livelihoods and poverty reduction policies.* Routledge, Oxon.

Eriksen, S., O'Brien, K., Rosentrater, L., 2008. *Climate change in Eastern and Southern Africa: impacts, vulnerability and adaptation.* Department of Sociology and Human Geography, University of Oslo, Oslo.

Few, R., 2003. Flooding, vulnerability and coping strategies: local responses to global threat. *Progress in Development Studies,* 3(1), 43-58.

Fidzani, B., Mlenga, W.S., Atlhopheng, M, Shatera, M.M., 1999. *Socio-economic effects of CBPP in NgamilandDivision of Agricultural Planning and Statistics,* Ministry of Agriculture, Gaborone.

Folke, C., 2001. *Socio-ecological resilience and behavioural responses.* Retrieved 08 March 2010: http://www.beijer.kva.se/publications/pdf-archive/Disc155.pdf .

Gumbricht, T., Wolski, P., Frost, P., & McCarthy, T. S., 2004. Forecasting the spatial extent of the annual flood in the Okavango delta, Botswana. *Journal of Hydrology,* 290(3-4), 178-191.

Hahn, M. B., Riederer, A.M., Foster, S.O., 2009. The livelihood vulnerability index: a pragmatic approach to assessing risks from climate change variability and change- a case study in Mozambique. *Global Environmental Change*, 19, 74-88.

Harry Oppenheimer Okavango Research Centre, 2010. HOORC data base. http://www.orc.ub.bw/datacat/mohembo.php?id=1anddatatype=_mon [Accessed 31 August 2010].

IPCC, 2008. Intergovernmental Panel on Climate Change Data Distribution Centre. <http://www.ipcc-data.org> (accessed 27.06.11).

Kgathi, D. L., Bendsen, H., Blaikie, P., Mbaiwa, J., Ngwenya, B.N., and Wilk, J., 2004. *Rural livelihoods, indigenous knowledge systems and political economy of access to natural resources in the Okavango Delta,* Botswana. Harry Oppenheimer Okavango Research Centre, University of Botswana, Maun.

Kgathi, D. L., Mmopelwa, G., & Mosepele, K., 2005. Natural resources assessment in the Okavango Delta, Botswana: Case studies of some key resources. *Natural Resources Forum,* 29(1), 70-81.

Kgathi, D. L., Ngwenya, B.N., Wilk, J., 2007. Shocks and rural livelihoods in the Okavango Delta. *Development Southern Africa,* 24(2), 289-308.

Leary, N., Adejuwon, J., Barros, V., Batima, P., Biagini, Burton, I., Chinvanno, S., Cruz, R., Dabi, D., de Comarmond, A., Dougherty, B., Dube, P., Githeko, A., Hadid, A., Hellmuth, M., Kangalawe, R., Kulkarni, J., Kumar, M., Lasco, R., Mataki, M., Medany, M., Mohsen, M., Nagy, G., Njie, M., Nkomo, J., Nyong, A., Osman-Elasha, B., Sanjak., E., Seiler, R., Taylor, M., Travasso, M., von Maltitz, G., Wadinga, S., Webbe, M., 2008. A stitch in time: general lessons from specific cases. In N. Leary, Adejuwon, J., Barros, V.,Burton, I.,Kulkarni, J., Lasco, R. (Ed.), *Climate change and adaptation. Earthscan,* London.

Magole, L., 2009. The "shrinking commons" in the Lake Ngami grasslands, Botswana: the impact of national rangeland policy. *Development Southern Africa,* 26(4), 611 - 626.

Magole, L., Magole, L.I., 2009. The Okavango: Whose Delta is it? *Physics and Chemistry of the Earth,* 34, 874-880.

Magole, L., Thapelo, K., 2005. The impact of extreme flooding of the Okavango River on the livelihood of molapo farming community of Tubu village, Ngamiland District, Botswana. *Botswana Notes and Records,* 37, 125-137.

Malala, B., 2009. *Report on floods in the Okavango.* District Commissioner's office, Maun.

Mazvimavi, D., Motsholapheko, M.R., 2008. Water resource use and challenges for river basin management along the ephemeral Boteti River, Botswana. In E. Manzungu (Ed.), *Towards a new water creed: water management, governance and livelihoods in southern Africa.* Rackwitz, Germany: InWEnt.

Mbaiwa, J. E., 2002. The socio-economic and environmental impacts of tourism in the Okavango Delta, Botswana: baseline study. *Harry Oppenheimer Okavango Research Centre,* University of Botswana, Maun.

McCarthy, J. M., Gumbricht, T., McCarthy, T., Frost, P., Wessels, K., Seidel, F., 2003. Flooding patterns of the Okavango wetland Botswana between 1972 and 2000. *Ambio,* 32(7), 453-457.

Moser, C. A., Kalton, G., 1971. Survey methods in social investigation. *Heinemann Educational Books*, England.

Niang, I., Nyong, A., Clark, B., Desanker, P., Din, N., Jalludin, M., Osman, B., 2007. Vulnerability, impacts and adaptation to climate change. In L. Otter, Olago, D.O., Niang, L. (Ed.), *Global change processes and impacts in Africa (*pp. 226-249). East African Educational Publishers, Nairobi.

Odero, K. K., 2006. Extending the sustainable livelihoods framework. *Discovery and Innovation,* 18(2), 83-91.

Reardon, T., 1997. Using evidence of household income diversification to inform study of the rural nonfarm labor market in Africa. *World Development,* 25(5), 735-747.

Schipper, E. L. F., Burton, I. (Eds.), 2009. Understanding adaptation: origins, concepts, practice and policy. In: E. L. F. Schipper, Burton, Ian (Eds.), *Adaptation to climate change* (pp. 1-8). Earthscan, London..

Scoones, I., 1998 Sustainable rural livelihoods: a framework for analysis. *IDS working paper 72.* Institute of Development Studies, London.

Scudder, T., Manley, R.E., Coley, R.W., Davis, R.K., Green, J., Howard, G.W., Lawry, S.W., Martz, D., Rogers, P.P., Taylor, A.R.D., Turner, S.D., White, G.F., Wright, E.P., 1993. *The IUCN review of the Southern Okavango Integrated Water Development Project.* IUCN, Switzerland.

Silitshena, R.M.K., 1982. Population movements and settlements in contemporary Botswana. *Botswana Notes and Records,* IV, 31-45

Smithers, J., Smit, B., 2009. Human adaptation to climatic variability and change. In: E. Lisa, Schipper, F., Burton, I. (Eds.*), Adaptation to climate change.* Earthscan, London.

Stone, W., 2001. Measuring social capital: towards a theoretically informed measurement framework for researching social capital in family and community life. Research paper no. 24. *Australian Institute of Family Studies,* Melbourne.

Swatuk, L.A., Motsholapheko, M., 2008. Communicating integrated water resources management : from global discourse to local practice -chronicling an experience from the Boteti River sub-Basin, Botswana. *Physics and Chemistry of the Earth,* 33(2008), 881-888.

Tlou, T., 1985. A history of Ngamiland 1750 to 1906- the formation of an African state. *Macmillan Botswana Publishing Company* (Pty) Ltd, Gaborone.

Turner II, B.L., Karsperson, R.E., Matson, P.A., McCarthy, J.J., Corell, R.W., Christensen, L., Eckley, N., Kasperson, J.X., Luers, A., Martello, M.L., Polsky, C., Pulsipher, A., Schiller, A., 2003. A framework for vulnerability analysis in sustainability science. *PNAS,* 100(14), 8074-8079.

Valentine, T.R., 1993. Drought, transfer entitlements and income distribution: the Botswana experience. *World Development*, 21(1), 109-126.

Vanderpost, C., 1991. Socio-economic constraints in molapo farming: baseline and monitoring surveys results 1987-1991. *Molapo Development Project, Department of Agricultural Research,* Ministry of Agriculture, Maun.

Vanderpost, C., 2006. Pathways of Human Sprawl in Wilderness Buffer Zones. *Population and Environment,* 27(3), 285-306.

Vanderpost, C. H. M., 2007. Human dimension of African climate and environmental change. In: L. Otter, Olago, D.O., Niang, L. (Ed.), *Global change processes and impacts in Africa* (pp. 72-92). East African Educational Publishers, Nairobi.

Wilk, J., Kgathi, D.L., 2007. Risk in the Okavango Delta in the face of social and environmental change. *GeoJournal,* 70, 121-132.

Wolski, P., Murray-Hudson, M., 2008. An investigation of permanent and transient changes in flood distribution and outflows in the Okavango Delta, Botswana. *Physics and Chemistry of the Earth,* 33, 157-164.

Wolski, P., Murray-Hudson, M., 2006. Recent changes in Xudum distributory of the Okavango Delta and Lake Ngami, Botswana. *South African Journal of Science,* 102, 173-175.

Woolcock, M., 2001. The place of social capital in understanding social and economic outcomes. *Isuma: Canadian Journal of Policy Research,* 2(1), 1-17.

World Bank, 1997. *Survey of living conditions: Uttar Pradesh and Bihar. Household questionnaire,* December 1997–March 1998.

In: Flooding
Editors: M. R. Motsholapheko and D. L. Kgathi

ISBN: 978-1-63117-607-4

Chapter 9

FACTORS INFLUENCING ADOPTION OF FLOOD RISK INFORMATION BY RESIDENTS IN FLOOD PRONE AREAS OF THE OKAVANGO DELTA, BOTSWANA

***O. T. Thakadu**[*1], **G. Mmopelwa**[2], **B. N. Ngwenya**[1],*
***B. Mathanjane**[2], **N. Munatsi**[2] and **W. V. Wanda**[2]*

[1]Okavango Research Institute
[2]University of Botswana

ABSTRACT

Flooding is a global phenomenon that continues to affect people socially, psychologically and economically and has been a concern for many years. Risk communication has been applied to inform the public about the potential harm caused by different environmental hazards. Literature has suggested that in cases where potential harm is perceived, the effectiveness of the risk communication interventions will depend on a number of factors such as perception of risk, trust, credibility and socio-demographic and cultural factors. The current study aimed at exploring factors that contribute to low adoption of flood risk warning information among local communities living in the Okavango Delta. The qualitative study, using a mixed method approach, was guided by the Mental Models and Trust Determination model to explore the socio-cultural and institutional environment within which flood risk communication took place, using a sample of 55 respondents affected by the floods. Findings revealed multiple factors such as a history of long residence in the area without experiencing floods, socio-cultural, myths, beliefs and perceptions towards water and floods, and low knowledge about floods and floods risks as factors contributing to the low adoption of flood risk warning information among the Delta communities. Others emanating from the sources of risk communitors include issues of trust and credibility. The study suggests that risk communicators and institutions should undertake preliminary synoptic audience assessments to guide communication interventions, participatory communication and land suitability and risk vulnerability before allocations.

[*] Corresponding author address: Private Bag 285. Maun. Botswana; Email: othakadu@ori.ub.bw

Keywords: floods, flood risk communication, flood risk, risk perception, Okavango Delta

INTRODUCTION

Environmental hazards such as earthquakes, tornadoes and floods have been a concern for many years worldwide (Wisner, Blackie, Cannon, & Davis 2004). Climate variability brings about issues of risks as disastrous environmental activities occur with changes in climate. In dangerous situations such as during tornadoes, floods and earthquakes, risk communication has been incorporated into communication processes to inform the public about the potential harm caused by such phenomena (Kunreuther, 1996). Literature has suggested that in case where potential harm is perceived, effectiveness of the risk communication interventions will depend mostly on the trust and confidence the recipient has towards risk communicators (e.g., Fessenden-Raden, Fitchen, & Heath, 1987; Weterings and Van Eijndhoven, 1989). If the risk communicator is viewed negatively by the target audience, adoption of information disseminated becomes challenging (Peters, Covello & McCallum, 1997).

Flooding is a global phenomenon that affects people socially, psychologically and economically (Kellens, Zaalberg, Neutens, Vanneuville & De Maeyer, 2011). Governments in most countries often offer assistance and early warning information to residents in flood prone areas (Brilly & Polic, 2005). The purpose of the early warning messages is to persuade people to take precautionary measures in preparation for floods. In Botswana, flooding incidences are becoming a common phenomenon, mainly the northern parts along the Okavango and Chobe rivers.

Settlements in the Okavango sub-District have experienced recurring floods since 2004 (Magole & Thapelo, 2006). The Delta receives water from the highlands of Angola, when it starts raining between October and April (Mendelsohn, VanderPost, Ramberg, Murray-Hudson, Wolski, & Mosepele, 2010). Flood level varies depending on the amount of rainfall recorded. In 2004, the Delta was submerged with floods, leaving about 30 homesteads and 40 crop fields damaged (Southern African Development Community, 2004). Both the residents and their livestock had to relocate to high ground.

In 2010, a number of villages in the Okavango sub-district were affected by the floods. The villages include Ikoga, Nxamasere, Etsha 13, Mohembo East, Kauxwi, Eretsha, Beetsha, Gudigwa, Jao Flats, Tubu and Nxaraga (Mosate, 2010). In an attempt to respond and manage disaster such as floods, the government has developed structures such as the National Disaster Management Office (NDMO) and the District Disaster Management Committee – DDMC (Government of Botswana, 2009).

Review of the literature shows that several studies have been conducted focusing on floods, impacts of floods, perception of flood risks and management, and risk communication (e.g., Baan & Klijn, 2004; Grothmann & Reusswig, 2006; Kellens, Zaalberg, Neutens, Vanneuville & De Maeyer, 2011; Plate, 2002). Lave and Lave (1991) found that literacy and science knowledge influence people's response to flood risks. Other factors such as location, experience with previous flood hazards and socio-demographic characteristic have been found to influence people's perception of risks, and adoption of risk communication (e.g., Baker, 1990; Fessenden-Raden, Fitchen, & Heath, 1987; Kellens, Zaalberg, Neutens,

Vanneuville, & De Maeyer, 2011). Socio-cultural and ethnic factors, knowledge about floods and the nature of the risk affect risk communication through their influence on the way the risk is framed, and consequently the risk response decision-making (Brilly & Polic, 2005; Fessenden-Raden et al., 1987; Vaughan, 1995). Owing to the multiplicity of factors that contribute to decision-making processes during risk situations such as floods, literature has established that risk communication can benefit from understanding the determinants of individual decisions that will lead to the required risk responses and decision processes (e.g., Chess, Salomone, & Hance, 1995; Terpstra, Lindell, & Gutteling, 2009).

Other studies have argued that effective risk communication should take into account information from the public and value the views and belief systems of the target communication audience (Trettin & Musham, 2000). While target audience factors may impede or facilitate adoption of risk communication, others (e.g., Peters et al., 1996; Fessenden-Raden et al., 1987; Vaughan, 1995; Trettin & Musham, 2000) have shown that risk communication is often overshadowed by issues of trust and credibility, thereby linking the communicator's expertise and trustworthiness to the effectiveness of risk communication. A study conducted in Botswana found that people were reluctant to relocate to safe places during floods as they did not trust the Government's motives for persuading them to move to such places (Magole & Thapelo, 2006). Moreover, the residents and the risk communicators' perception of floods differed, making risk communication even more challenging. According to Vaughan (1996), effective risk communication will be realised when the socio-cultural environment of the target audience is understood by the communicators. Regionally, few studies have been conducted on risk communication. A study by Nyathi (2013) focussed on understanding how indigenous knowledge systems (IKS) can be used in risk communication in southern Africa. The study concluded that IKS presented an untapped potential in risk communication. Other studies have emphasised on flood and risk management, together with local communities adaptation and coping strategies.

While the need for understanding the socio-cultural environments and factors that contributes to people's response to flood risk information has been advanced in literature, little empirical research has been conducted in Botswana to explore factors that will facilitate adoption of flood risk communication interventions among communities residing in flood prone areas. Much of the studies indicated have been conducted in socio-economic, cultural and political environments different from Botswana, and though they shed light on requisite factors communicators should consider when communicating risk, it remains important that socio-cultural environments within which risk communication takes place is clearly understood.

The recurrence of floods in northern Botswana and the resulting need to alert the communities living in flood prone areas warrant that risk communication intervention and approaches are guided by the empirical studies on the socio-cultural environments within which such communication is targeted. Since the recurrence of the floods in the Okavango Sub-District, government has responded by informing the residents in flood prone areas to relocate to safer locations (Mosate, 2010). Despite the attempts to communicate flood risks, the local communities seem not to respond to the early warnings as they continue to stay in such flood prone areas. Most risk communication interventions implemented within these communities draw from approaches, methods and models developed mainly in developed countries. These risk communication approaches may therefore not be applicable in developing world context, as they are implemented without guidance from any local empirical studies (Pe Benito Claudio, 1988).

Against this backdrop, the current study aimed at exploring factors that contribute to low adoption of risk warning information on floods by the local communities living in the Okavango Delta. The specific objectives are to; 1) identify the sources of flood risk communication available to the local residents; 2) examine the approaches used to communicate flood risk warning information to local residents; 3) determine local people's perceptions of risk information sources and the risk information disseminated; 4) assess the local people's knowledge about floods; 5) examine communicators' and local people's perceptions on floods and flood risks; and 6) identify challenges to risk communication in the Okavango Delta.

THEORETICAL FRAMEWORK

The study was guided by two theoretical models, the Mental Models and the Trust Determination Model, to investigate the factors that influence people's adoption of flood risk messages. The mental models approach asserts that people interpret a social phenomenon based on their affective cognition (Morgan, Fischhoff, Bostrom, & Atman, 2002). The model posit that for risk communication to be effective, it has to take into account the emotions, behaviours, beliefs, views and attitudes of the target audience. The approach is premised on the following; 1) the risk information recipients need to understand the exposure, outcomes and mitigation processes to facilitate making decisions about the risky situation, 2) the beliefs that the recipients hold affects how they view and make use of any new information and 3) the risk information should be represented appropriately and be reinforced with other aids (Atman, Bostrom, Fischhoff, & Morgan, 1994). After assessing the cognition processes of the general public, the risk communicator should then compare and contrast their own views, behaviours, attitudes and knowledge in order to identify areas of commonality and discrepancy. After identifying the major differences, interventions and risk communication messages should be designed in a manner that will address the beliefs, attitude and knowledge of the target audience. This is so because proper risk information based on the mental model's approach should be aimed at adding vital missing information and displacing misconceptions that impact greatly on decisions (Morgan et al., 2002). It should also be constructed to confront misperceptions about the reality of the situation.

The Trust Determination model on the other hand focuses on credibility and trust factors between risk information sources and the recipients. The framework posits that the process of risk communication can be made more effective in an environment of trust and credibility between communicators and recipients (Covello, Peters, Wojtecki, & Hyde, 2001).

The two factors will contribute to the way people perceive, process and respond to risk. Trust and credibility are in turn products of demonstrated caring and empathy, dedication and commitment, competence and expertise, as well as honesty and openness (Figure 1).

The Trust Determination Model suggests that the low adoption of flood risk information in the Okavango Delta might not be related to the choice of the channel and the content of the message, but to the trust and credibility that is assigned to the responsible disaster management institutions. Hence, effective risk communications requires the institutions to build a relationship of mutual trust. The two models therefore becomes relevant for the study as the mental models approach will help understand the target risk message audiences' beliefs, perceptions and attitudes towards the risks whereas the trust determination model will shed light on the way the risk communicators are perceived.

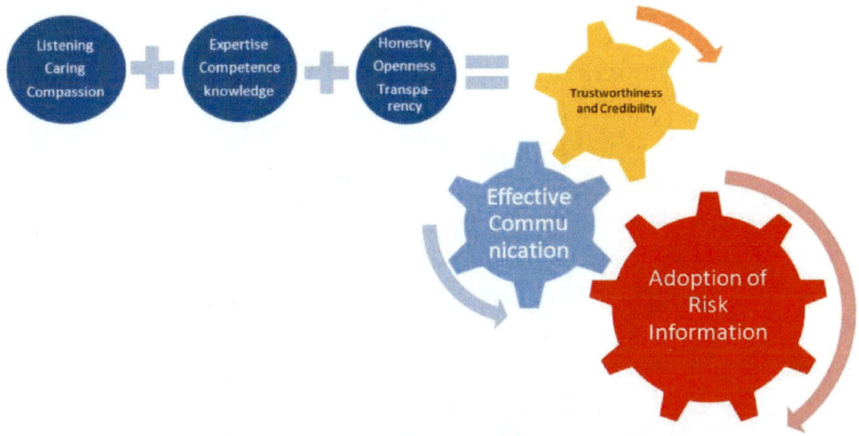

Figure 1. Trust Determination Model (adapted from Lundgren & McMakin, 2004; Slovic, 1999; Fessenden-Raden, Fitchen, & Heath, 1987.

METHODS

Study Area

The study was carried out in Ngamiland District, focussing in the Okavango sub-District, covering the villages of Etsha 13 and Ikoga (Figure 2).

Etsha 13 has a population of 2 377 while Ikoga has 673 (Central Statistics Office [CSO], 2012). The two villages were adversely affected by the floods in 2009 (District Disaster Management Committee, 2010). Most people in the Delta depend on natural resources found in the wetland, such as fishing and flood recession cultivation, which is commonly known in the Delta as *molapo* farming (Kgathi et al., 2006). Flooding in the Okavango Delta is mainly influenced by the amount of water inflows from the Angolan highlands.

Design and Sampling

The study followed a qualitative research paradigm. The exploratory nature of the study influenced the choice of the design, as it permits generation of information on a topic that has not been locally studied. The utility of the qualitative approach is that it does not confine researchers in their search for knowledge (Ary, Jacobs, Razaveih, & Sorensen, 2006).

The non-probabilistic procedures of purposive and snowballing sampling methods were used to select respondents. The respondents were pre-defined prior to data collection as those who were privy to relevant information on flood risk and communication.

These included people who had been affected by the floods in the villages of Etsha 13 and Ikoga, as well as key informants and opinion leaders. Key informants included those involved in risk communication, being; 1) the District Officers; 2) the village development committee (VDC) chairpersons and 3) *Dikgosi* (chiefs, singular *Kgosi*) in the study areas. These constitute part of the District Disaster Management Committee (DDMC) membership and play a key and active role during risk communication.

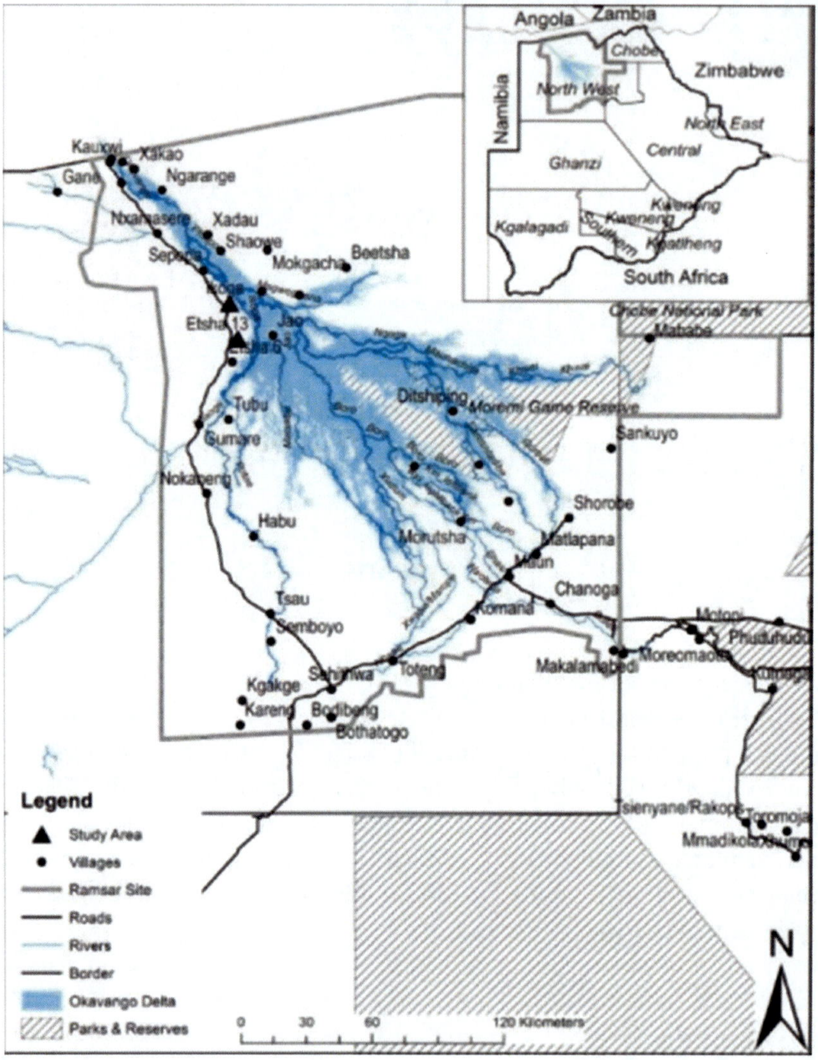

Figure 2. Map showing study area.

Instruments and Procedures

A semi-structured interview guide was developed, guided by the objectives of the study. When constructing the interview guide, some questions stated in the literature were considered, such as the knowledge of respondents and those related to trust and credibility (Fischoff, Bostrom & Quadrel, 1993; Peters, Covello & McCallum, 1997). There were two sets of interview guides, one for the general respondents affected by flooding and the other for the key informants.

The respondents' interview guide captured the demographic characteristics, perceptions and knowledge and the sources of risk information. The key informants' interview guide captured information on the approaches used in flood risk communication, frequency of the communication, what is being communicated to the local people, how indigenous knowledge of the local people is factored in the communication process, and the challenges they faced during risk communication. The interview guide was mainly comprised of open-ended questions.

The interview guides were pre-tested in Nxaraga, a village previously affected by floods. The pre-testing exercise was done to find out if the questions were clear and understandable to the respondents. Interviews were recorded using digital recorders and later transcribed. Administering a semi-structured interview, Focus Groups discussion (FGD) and participant observations were used during data collection. Two FGDs were conducted, one in Etsha 13 with six individuals and the other in Ikoga village, attended by five people.

Six semi-structured interviews were made with key informants and opinion leaders comprising of village headmen (*Kgosana*), VDC members and a government official. Three key informants were from Etsha 13, two from Ikoga and one was a government official responsible for disaster management under the DDMC. Only one of the key informants was a female, with the remaining five being males.

Data Analysis

Primary data were analysed using qualitative techniques. Thematic analysis was used to identify the major themes coming from the ideas and phrases expressed by respondents.

Data were organised into categories that summarised and brought meaning to the text (Ary et al., 2006). The identification of these major themes helped in understanding the respondents' affective and cognitive perception towards risk warning information. Data were finally interpreted based on the meaning of relationships between information within and between categories, what is important and what was learnt about the respondents (Powell & Renner, 2003).

RESULTS AND DISCUSSION

Demographics

A total of 55 respondents were interviewed. Forty of the respondents were individuals who had been directly affected by floods. Two-thirds of the respondents were females. Literacy was low within the study area, with 63% (n = 25) of the respondents without any formal education. The mean age of the respondents was 49.76 years (Standard deviation = 16.05), with a modal age range of 50-59 accounting for almost 32% of the respondents (n = 12) (see Table 1). The other respondents went through primary education (10%), while a quarter had gone through junior certificate level. The high illiteracy level in the study area suggests that communicating flood risks among these communities may be challenging. Studies have shown that the level of education (literacy) influences responses to risk situations and information.

Table 1. Respondents' age groups

Age group (years)	Total (n)	%
20-29	5	13.2
30-39	6	15.8
40-49	5	13.2
50-59	12	31.6
60-69	3	7.9
70+	7	18.4

The respondents were predominantly Wayei (60%), followed by Hambukushu (37.5%) and Basubiya (3%). As suggested in literature, the multiplicity of ethnic groups represent different perceptions and beliefs towards flood risks in the area (Fessenden-Raden et al., 1987; Vaughan, 1995) the findings also revealed that the origin of these ethnic groups also contributed to their level of knowledge regarding floods and their associated risks. The three ethnic groups are often referred to as *BaNoka* translated "river people", indicating the ethnic groups' intimate relationships with the river (Tlou, 1985; Magole & Thapelo, 2005). Seventy-two percent of the respondents indicated that their main livelihood activity was *molapo* farming and the remaining percentage shared among other sources of income such as small business enterprises, fishing, basket weaving, cash employment and beer production. The reliance upon *molapo* farming also demonstrates the communities' attachment to the riverine environments, a scenario that may be a challenge to risk communication. Any attempt to relocate the communities may be like cutting them from their mainstay livelihood source.

Sources of Risk Communication

Flood risk information in Etsha 13 and Ikoga is communicated by the District Disaster Management Committee (DDMC). The District Officer Development (DOD) of Gumare said that they do communicate the flood risk with communities residing in flood prone areas. The DDMC is divided into sub-committees which are assigned specific tasks in disaster management. For example, the Public Awareness (PA) Team is responsible for disseminating risk information to the public. It comprises the Village Development Committee (VDC) members, *Dikgosi* and their subordinates, the Department of Water Affairs and Red Cross. The Department of Water Affairs informs the DDMC on the water situation. The information is then passed on to the communities. Other stakeholders include the politicians such as the councillors and area member of parliament.

The composition of risk communication team (PA) demonstrates concerted and coordinated efforts. The team effort approach to flood risk communication is commendable as it overcomes the likelihood of giving communities conflicting and contradictory information about the risk (Fessenden-Raden et al., 1987). The Government of Botswana has developed a National Disaster Risk Management Plan (Government of Botswana, 2009) in 2009 to guide and inform processes, procedures and practice during disasters. The plan also outlines the different structures operating nationally, regionally and locally to ensure proper coordination, management and monitoring of disaster events. The plan notes that early warning information is coordinated from one centralized authority so as to ensure the accuracy of the information flowing between the risk communicators and the community. During flood risk communication, the risk communicators are thereby guided by the National Disaster Risk Management Plan.

Approaches and Channels Used in Risk Communication

Risk communicators conduct house to house campaigns and use public address system to inform the general public about the impending risks. They also hold *Kgotla* meetings where the *Kgosi*, District Commissioner and other government officials address the communities on matters pertaining to floods and flood risks. Radios are also used to spread information on flood risk. The flood risk communication is meant to inform the local people about the floods and demand people residing near the river to evacuate to high and safe land. The respondents indicated the following as channels through which they received flood risk information; house to house, radio, *Kgotla* and public address system. These represent a multi-mix approach that utilizes a diversity of channels and is good for broad-base impacts in terms of reach and target audience (Thakadu & Tau, 2012).

Though the National Disaster Risk Management Plan of 2009 indicated that short-text message system (SMS) and television were used to communicate risk to people at risk, none of the respondents in the study area indicated that they ever received the message through this means. The National Disaster Risk Management Plan stated that "communication among various sectors and institutional level in the country will be carried out using various communication mediums such as the telephone, fax and internet". These mediums are used to communicate high risk information during cases of severe disaster. In the two study areas, services such as electricity and telephones are limited. Both the *Kgotla*[1] offices at Ikoga and Etsha 13 villages do not have electricity hence cannot access the internet and fax. This might lead to risk communication being delayed.

Perception of the Sources and Risk Information

Timing and Frequency of Risk Communication

According to the key informants, floods often start between January and June every year. The DDMC through the PA team starts to sensitize people about impending floods at the beginning of the flooding month. Risk warning information is communicated at least thrice each month. However, respondents stated that risk information comes to them late. They said risk communicators start to sensitize them on floods only when the flood waters are almost within homesteads. Some of the respondents said that they never heard flood warning

[1] *Kgotla* refers to both an institution and a place. As an institution, *Kgotla* is the decision-making body at the village or community level; as a place, it serves to accommodate a wide range of fora such as communal discussions, deliberations, and public information dissemination on issues of mutual interest to the broader community.

messages or information. Some said they were shocked to find water surrounding their homestead. This forced them to seek for help from the authorities.

Credibility of Information and Its Sources

Most of the respondents said that the floods they observed were the first during their lifetime. Based on their history and background in the area, they found it difficult to believe the sources because most have not seen floods before. Most of the people took it as fairy-tales. These findings are supported by Panwher (2010) who found that since the Indus river did not experience heavy floods for a long time, people tend to end up taking it that it will no longer flood. Communicating flood risk information with the Hambukushu and Wayei is a challenging task because they have lived along river courses for a long time without any danger.

As much as these people lived many years without experiencing any floods in their areas, the other complexity with the area is that even though intentions to evacuate may be there amongst residents, the risk communicators are not able to specify the exact flow directions of the flood waters. This leaves the whole intervention to look like guess work. Some of the respondents said that the risk communicators once told them that water will reach their area at some point in time but the water never came. When the communicators returned a year later with the same warning message, they did not take them seriously as they considered it "a familiar threat, which never materializes". Failure by responsible authorities to accurately predict flood flows end up making local communities to lose trust and confidence on the risk communicators as they doubt their knowledge and expertise. According to Peters et al. (1997), when people question communicators' knowledge and expertise, it results in declines of trust in public institutions and credibility. When people do not trust the risk communicators, they end up being reluctant to evacuate.

Relying on the experience of long residence without flood events, residents of Ikoga and Etsha 13 did not immediately evacuate after being told that the floods were approaching. Respondents in focus group discussions said they waited for the water to approach their homes as they did not trust government officials. This showed the lack of trust people had on the risk informants. It follows that the people moved when it was already late to save their property. In some instances some people, who had earlier been relocated, returned to their original plots after the floods receded, hence being affected again in the next flooding season.

Resettling in flood prone areas by flood victims after the floods showed that people were willing to take further risks, or believed that the floods will not recur. This may be so because the livelihood activities of these communities are closely linked to the river and its resources. The communities practice traditional *molapo* farming and harvest natural resources from the riverine environment. This may explain why people are willing to take risks by resettling in former flooded areas. Permanent relocation to dry areas may be considered detachment from their sources of livelihood. It is this deep rooted reliance on riverine resources for livelihoods that complicates and confounds flood risk communication.

Local People's Knowledge about Floods

The findings indicated that the respondents did not know much about floods, their origins and causes, and the measures to take to protect themselves and their property from the potential destruction by floods. Lave and Lave (1991) also found that some communities in the United States of America had little knowledge about floods. The community of Ikoga and Etsha 13 were only aware of one river channel which has been flowing seasonally, without posing danger to their lives and livelihood activities. Focus group discussion participants corroborated this when they said that they have stayed in the area for a long time without experiencing any flood. According to them, things changed from 2008 when the river started flowing in what they came to know as 'old channels' that they were even not aware of. The respondents said they were not aware that their homesteads were located in a floodplain. When they were warned of the impending floods by the DDMC, they did not take the information seriously as they had neither experienced floods nor aware that they had settled in flood prone areas.

Since most of them never went to school, this may be a possible explanation as to why most did not know where the floods come from. Lave and Lave (1991) findings indicated that knowledge regarding floods is influenced by level of literacy. It is therefore not surprising that the respondents were not knowledgeable on the floods and their origins as they were mostly illiterate.

Lack of awareness of residency in floodplains and illiteracy may have played a role in respondents' reluctance to evacuate and to take heed of the flood risk information disseminated by the risk communicators. Whereas the Wayei respondents did not know where the floods came from, the Hambukushu respondents, who originally migrated from Angola to Botswana, rightly stated that the floods come from Angola, emanating from rains during rainy seasons. They said that since they stayed in Botswana they did not know that they settled in floodplains or along channels.

Sixty percent of the respondents indicated that they did not know how floods came about, indicating that they have not been exposed to any information pertaining to floods, nor have they been taught about floods. The Botswana basic 7-10 years school syllabus covers topics on floods and the Okavango Delta. However, most of the respondents never went to school (63%, n = 25), and hence did not know much about the floods. According to Fessenden-Raden et al. (1987), when people lack basic concepts of a phenomenon, the potential to understand information presented to them is reduced since the concepts presented become more abstract to relate with and act responsibly.

Perception of Floods and Flood Risks

Risk Communicators' Perception about Floods and Flood Risks

Risk communicators perceived floods to be a natural phenomenon that occurred due to climate variability. They also opined that the negative impacts of floods can be reduced because floods can be forecasted. Their main purpose is to ensure that people are informed about the floods and their dangers in order that they relocate on time. They also perceived floods to be destructive. The risk communicators also indicated that floods are capable of damaging the local people's houses and other household goods. Livestock may also be endangered. According to them, the flood comes with dangerous water animals such as crocodiles, hippos and snakes which may pose danger to the local people.

Local People's Perception of Floods and Flood Risks

Respondents in Etsha 13, especially from the Wayei ethnic group, held a myth that "someone broke a dam upstream and then water started gushing out", resulting in floods which negatively affect their livelihood. Some said that "whites opened a water source in some place upstream hence filling up our land". Focus group discussants substantiated this statement that floods are man-made saying "someone blasted a rock in Namibia". These statements are indicative that people perceive floods and the resulting damages to be man-made. The findings are similar to those noted by Lave and Lave (1991) who found that people tend to pass blame to someone or something. Some respondents even blamed the government for water coming their way. They implicated that government had blocked the river channels with sand at some point in the past and the blockage diverted water towards them.

Some said that they believed that water, and consequently floods, come from God. Others held views that floods are a natural phenomenon and could occur anytime. There were few respondents who indicated that flood water comes from the ancestors. One of the key informants, Etsha-13 VDC Chairperson indicated that people in his community have beliefs, myths and perceptions regarding the floods. He said "water is our totem, hence we don't dread it, we revere it". He further said that if water happens to come their way in the form of floods, they move temporarily, not because they are afraid but just to give way. "When it recedes we go back", he said. This was corroborated by Ikoga village headman when he reiterated that water is part of their livelihood and culture; hence people have to learn to live with it.

In most instances, people did not evacuate immediately when they were informed that floods were coming. They adopted a wait and see attitude even though they were warned about the danger of attacks by aquatic animals including snakes, crocodiles and hippos. After the destruction caused by the floods to their homes, they perceived floods as a catastrophe that leaves people homeless and in poverty. Even among some, the belief that water is a totem seems to be changing because after the floods destroyed property and forced people out of their homesteads, they did not want to be associated with water.

Risk Communication Challenges

Both the risk communicators and the respondents expressed some concerns on constraints to risk communication among the communities. Some respondents indicated that in resettlement camps, a number of challenges exist which tend to work against people's willingness to move to relocate to the camps. For example, the respondents complained about the insanitary conditions and frequent theft. There were also reports of political interference whereby the opposition politicians influenced people not to relocate unless residential structures were built for them and/or be given the same treatment as the 2009 flood victims. They informed the affected people not to take heed of the warning messages until government gave them a fair deal as victims.

In their endeavour to communicate risk to the locals, risk communicators often face challenges on the efficiency of risk communication. Some of the factors that impede the effectiveness of risk communicators in the area included inadequate resources, time constraints and the vastness of the area which risk communicators had to cover. For risk communication to be effective there has to be enough resources to facilitate efficient information dissemination. Resource constraints in the Okavango sub-district, mainly human and other key resources, have become a big challenge for the risk communicators, and often complicated the risk communication process.

Risk communicators often had limited time to communicate their messages in any given area and cover all the affected areas within reasonable time prior to the floods. As a result, information reached some areas late hence the late evacuation of some affected people. According to the Africa Regional Strategy for Disaster Risk Reduction (African Union/New Partnership for Africa's Development, 2004), for the local people to make informed decisions during a disaster, risk warning information should be readily available and given timely. However, in the study area, there is a gap between risk communication and emergency response due to the shortage of resources. The main challenge in communicating flood risk in these areas was that the risk communicators were not certain on areas which will be affected because the channels shift and the water keeps changing courses – a common phenomenon in the Okavango Delta. Consequently, the communicators tended to concentrate their information dissemination efforts on households resident along river banks.

Conclusion

The main objective of the study was to determine factors that contribute to the adoption of flood risk warning information. Findings have revealed multiple factors that may be responsible for low adoption of flood risk warning information among communities of Etsha-13 and Ikoga. The factors include a) long history of residence in the area without experiencing floods, b) myths, beliefs and perceptions towards water and floods, and c) low knowledge about floods and floods risks. Socio-cultural factors also seem to play a role in that different ethnic groups differed in perception and knowledge about the floods and their origins.

High levels of illiteracy may also be contributing to low levels of knowledge regarding floods because most of the respondents never went to school to learn basic concepts regarding floods. Elements of (mis)trust and lack of credibility of the risk communicators were also indicated. This comes about due to the uncertainties in forecasting the magnitude and the direction of floods, compounded by the dynamic flow patterns of the channels and the distribution of the water within the Okavango River and its channels. The other factor that may be confounding risk communication and potential relocation to safe places is the intimate connection of the communities with their environment, mainly the Okavango Delta. The Okavango Delta and its waters serve as a water, fishing, *molapo* farming activity and a host of other activities that support household livelihoods.

The study also found multiple sources, channels and approaches used in risk communication. The communicators used team approach in communicating with communities at risk. The team was composed of different expertise, including representation from traditional and local institutions. Albeit this, there were challenges such as inadequate resources for risk communicators to effectively disseminate risk messages and respond to floods and flood risks.

Additionally, the conditions at the resettlement camps were not attractive enough to facilitate further relocations. The quests to gaining political mileage by politicians who influence people not to relocate unless certain demands are met by the government are also issues of concern. On the other hand, government should be cautious not to set double-standards on benefits given to people who were relocated because this may further complicate risk communication in the area.

On the forgoing, the study suggests the following to improve the practice of risk communication:

- Owing to the big gap between what is known by risk communicators and the local communities, there is a need for risk communicators to adopt the mental models approach to guide their risk communication interventions. The approach advances the need to undertake target audience assessment in regard to beliefs, myths and perceptions and designing interventions that can address themental models of the target audience. By so they may effectively alter the mental models of the information recipients and facilitate more informed and quicker decision-making.
- The land authorities (Land Board) should undertake a thorough assessment of land suitability before residential plot allocations are made. Areas prone to floods and flooding should be avoided when it comes to plot allocations, mainly for residential use.
- Risk communication can be improved by adopting participatory communication approaches which will accommodate people's knowledge and input in risk planning and management. Traditional adaptive approaches can be blended with conventional approaches for effective relief. The approach may further help repair the trust and credibility issues among risk communicators.
- Floods and flood risks public awareness campaigns to educate people about floods should be a continuous process, as opposed to the current interventions targeting flood seasons. This will help to ensure that people get well informed about floods and the associated potential risk well on time and in large numbers.
- Land that is prone to any natural disaster should not be allocated to people. Elderly people can also be incorporated into this exercise as they know more about the villages they live in. There should be flood awareness campaigns to educate people about floods, their origin and the capacity of their danger and damage to people's lives. This is because local people do not know about floods, and they also underestimate the threat of flooding.
- Adequate resources should be dedicated to risk communication teams to make risk communication more efficient and effective. Such resources should include human and financial resources.

Studies such as this one should be undertaken to understand how risk communication can be done most effectively in different socio-cultural environments. Communicating risk, like all other forms and domains of communication, needs understanding of the target audience and analysis of communication behaviours, knowledge and perception of risk events. This may differ across societies, cultures, communities and generations.

REFERENCES

African Union/New Partnership for Africa's Development. (2004). Africa Regional Strategy for disaster risk reduction. *Disaster Risk Reduction for Sustainable Development in Africa*. 3 - 8.

Ary, D., Jacobs, L. C., Razaveih, A., & Sorensen, C. (2006). *Introduction to research in education*. Belmont, CA: Thomson Wadworth.

Atman, C. J, Bostrom, A, Fischhoff, B., & Morgan, M. G. (1994). Designing risk communications: Completing and correcting mental models of hazardous processes. *Risk Analysis, 14*(5), 779-788. doi:10.1111/j.1539-6924.1994.tb00289.x

Baan, P. J. A., & Klijn, F. (2004). Flood risk perception and implications for flood risk management in the Netherlands. *International Journal of River Basin Management, 2*(2), 113-122. doi:10.1080/15715124.2004.9635226.

Baker, F. (1990). Risk Communication about environmental hazards. *Journal of Public Health, 11*, 341-359.

Bodgan, B. C., & Biklen, S. K. (1998). *Qualitative research in education: An introductory to theory and methods*. Needham Heights, MA: Allyn and Bacon.

Brilly, M., & Polic, M. (2005). Public perception of flood risks, flood forecasting and mitigation. *Nat. Hazards Earth Syst. Sci. 5*(3), 345-355. doi:10.5194/nhess-5-345-2005

Central Statistics Office [CSO]. (2012*). 2011 population and housing census. Population villages and associated localities*. Central Statistics Office. Gaborone: Government Printer.

Chess, C., Salomone, K. L., & Hance, B. J. (1995). Improving risk communication in government: Research priorities. *Risk Analysis, 15*(2), 127-135.

Covello, V., Peters, R., Wojtecki, J., & Hyde, R. (2001). Risk communication, the west Nile virus epidemic, and bioterrorism: Responding to the communication challenges posed by the intentional or unintentional release of a pathogen in an urban setting. *Journal of Urban Health, 78*(2), 382-391. doi:10.1093/jurban/78.2.382

District Disaster Management Committee (2010). *Okavango Sub-District Disaster Report*. Maun: Northwest District Council

Fessenden-Raden, J., Fitchen, J. M., & Heath, J. S. (1987). Providing risk information in communities: Factors influencing what is heard and accepted. *Science, Technology, & Human Values, 12*(3/4), 94-101. doi:10.2307/689388

Fishoff, B., Quadrel, M. J. (1993). Risk Perception and Communication. *Annual Reviews, 14*, 183-203.

Food Agriculture and Natural Resources. (2004). *Regional flood watch*. Harare: SADC Regional Remote Sensing Unit.

Government of Botswana. (2009). National Disaster Risk Management Plan. National Disaster Management Office. Gaborone.

Grothmann, T., Reusswig, F. (2006). People at risk of flooding: Why some residents take precautionary action while others do not. *Natural Hazards 38*(1-2), 101-120, doi:10.1007/s11069-005-8604-6.

Harris, F. (2004). *Global Environmental Issues*. West Sussex: John Wiley.

Hayenhjelm, M. (2006). Assymetrics in risk communication. *Risk Management*, 1-15.

Kellens, W., Zaalberg, R., Neutens T, Vanneuville, W., De Maeyer, P. (2011) An analysis of the public perception of flood risk on the Belgian coast. *Risk Analysis 31*(7): 1055-1068; doi:10.1111/j.1539-6924.2010.01571.x

Kgathi, D. L., Kniveton, D., Ringrose, S., Turton, A. R., Vanderpost, C. H. M., Lundquist, J., Seely, M. (2006). The Okavango; a river supporting its people, environment and economic development. *Journal of Hydrology*, *331*(1-2), 3-17.

Lave, T. R., Lave, L.B. (1991). Public perception of risks of floods: Implications for communication. *Risk Analysis*, 255-266.

Lundgren, R., McMakin, H.A. (2004). Risk communication: A handbook for communicating environmental, safety, and health risks. Colombus, Ohio: Battelle press.

Mendelsohn, J., VanderPost, C., Ramberg, L., Murray-Hudson, M., Wolski, P., Mosepele, K. (2010) *Okavango Delta: Floods of life*, Windhoek: RAISON.

Morgan, M. G., Fischhoff, B., Bostrom, A., Atman, C. J. (2002). *Risk communication: A mental models approach.* Cambridge: Cambridge University Press

Mosate, M. (2010). *An update on the floods situation.* Gaborone: Botswana Government Printer.

Nyathi, M. (2013). Disaster risk communication: A dichotomous approach incorporating indigenous knowledge systems. *Journal of the Eastern and Southern Africa Regional Branch of the International Council on Archives.* 32, 30-44.

Panhwer, S. A. (2010). Indus River flood: The fate of squatter settlements and adjacent villages, 1-2.

Pe Benito Claudi, C. (1988). Risk analysis in developing countries. *Risk Analysis, 8*(4), 475-478.

Peters, R. G., Covello, V. T., McCallum, D. B. (1997). The determinants of trust and credibility in environmental risk communication: An empirical study. *Risk Analysis, 17*(1), 43-54. doi:10.1111/j.1539-6924.1997.tb00842.x

Plate, E. J. (2002). Flood risk and flood management. *Journal of Hydrology 267*(1â€"2): 2-11, doi:org/10.1016/S0022-1694(02)00135-X

Powell, E.T., Renner, M. (2003). Analysing qualitative data. Wisconsin: Board of Regents of University of Wisconsin system.

Southern African Development Community (2004). Regional flood watch: February season 2003/2004. Gaborone, SADC FANR Regional Early Warning System. Retrieved from http://www.fews.net/docs/Publications/1000404.pdf

Terpstra, T., Lindell, M. K., Gutteling, J. M. (2009). Does communicating (flood) risk affect (flood) risk perceptions? Results of a quasi-experimental study. *Risk Analysis, 29*(8), 1141-1155. doi:10.1111/j.1539-6924.2009.01252.x

Thakadu , O.T., Tau, O.S. (2012). Communicating environment in the Okavango Delta, Botswana: An exploratory assessment of the sources, channels, and approaches used among Delta communities. *Science Communication*,

Kunreuther, H. (1996) Mitigating disaster losses through insurance. *Journal of Risk and Uncertainty, 12*(2-3), 171-187. doi:10.1007/bf00055792

Magole, L. Thapelo, L. M. (2006). The impact of extreme flooding of the Okavango River on the livelihood of the *molapo* farming community of Tubu village: Ngamiland Sub-District, Botswana. *Botswana Notes & Records*, *37*, 125-137.

Tlou, T.C. (1985). A history of Ngamiland. Gaborone: Macmillan Botswana Publishing.

Trettin, L, Musham, C. (2000). Is trust a realistic goal of environmental risk communication? *Environment and Behavior, 32*(3), 410-426. doi:10.1177/00139160021972595

Vaughan, E. (1995). The significance of socio-economic and ethnic diversity for the risk communication process. *Risk Analysis, 15*(2), 169-180. doi:10.1111/j.1539-6924.1995.tb00311.x

Weterings, R. A. P. M., & Van Eijndhoven, J. C. M. (1989). Informing the public about uncertain risks. *Risk Analysis, 9*(4), 473-482. doi:410.1111/j.1539-6924.1989.tb01258.x

Wisner, B., Blackie, P., Cannon, T., & Davis, I (2004). *At risk: natural hazards, people's vulnerability and disasters*. New York. Routledge.

In: Flooding
Editors: M. R. Motsholapheko and D. L. Kgathi
ISBN: 978-1-63117-607-4

Chapter 10

FISH, FLOODS AND LIVELIHOODS IN THE BOTETI RIVER, BOTSWANA

K. Mosepele*

Okavango Research Centre, University of Botswana

ABSTRACT

Fish is a major source of livelihoods for riparian communities. It contributes significantly to food and nutrition security, rural employment and general poverty alleviation. Studying the dynamics of this resource in riverine/flood pulsed systems provides critical information essential for management of this resource. It is also important to understand the impact of extreme flood events on this resource, so that their impact on the rural livelihoods can be established. In the absence of comprehensive data linking the effects of extreme floods to riverine fish dynamics, predictions can be made using holistic environmental flow assessment (EFA) methodologies. Therefore, this approach was used to assess the effect of extreme floods on the Boteti River fishery. Typical of holistic methods in EFA methodologies, baseline information on the biology and ecology of Boteti River fisheries were determined. Fish data were collected using standardised experimental fishing methods, while socio-economic data were collected using structured interviews. Various fish community indices were used to assess spatio-temporal variations in fish community structure and feeding ecology of selected species. Two fish indicators, based on fish guilds, were then used to predict the effect of extreme flooding on fish populations. Qualitatively, results revealed that there were spatio-temporal variations in fish community structure in the river systems, though there were no significant differences ($p>0.05$) in selected indices among the different sites. However, there were significant differences ($p<0.05$) in relative abundances and morphometrics of selected species among sites in the river system. This suggests that variability in the fish community is observed at the species level and not the population level. Results also revealed that the terrestrial environment is a major source of energy for the river's fish community, driven by flooding. Understandably, the fish resource in this river is a major source of livelihoods for the riparian community, though fishers had diverse economic activities. It was predicted that while extreme floods had negative impacts on fish populations, prolonged flooding also contributes to increased fish

* Author address: P/Bag 285, Maun. Botswana. Telephone: +267 686 1833; E-mail: kmosepele@ori.ub.bw

production for some species. While negative impacts on fish production would have a deleterious impact on fishers, it was also concluded that their diverse economic activities is an adaptation strategy to variability in fish availability. Another major conclusion is that the Boteti River fishery is very resilient (and dynamic), and will always return to normal after any extreme event such as drought or flood.

Keywords: environmental flows, flood pulse, flood risk, Okavango

INTRODUCTION

Rivers provide fish as a major source of livelihoods for most (socio-economically marginalised) riparian communities (Petrere et al., 1998; Allan et al., 2005; Dugan et al., 2010). Out of the 51 million fishers in the world, only 500,000 of them do not work in small scale fisheries (Berkes, 2003). These small scale fisheries (ssf) are relatively profitable, are resilient to shocks and crises, and contribute significantly to poverty alleviation and food security (Allan et al., 2005; FAO, 2005). In some countries (e.g., Uganda, Cambodia, Bangladesh), inland fisheries contribute over 60% of animal protein (Welcomme, 2011). Inland fisheries are critically important in landlocked countries where they are a major source of protein to impoverished communities (Dugan et al. 2010). While the monetary value of these inland fisheries can be quite substantial (Dugan et al. 2010), the smaller sized species in these assemblages are a major source of micro-nutrients (e.g., Vitamin A, Iron, Zinc, etc.) that are essential to women and children (Larsen et al., 2000; Roos et al., 2008). Fish is also an important source of protein, especially in poor countries (Delgado et al., 2003). Essentially, the effects of food insecurity in Sub-Saharan Africa (Smith et al., 2006) can be ameliorated through a judicious exploitation of inland fisheries.

Fish production in these systems is explained through Junk et al.'s (1989) flood pulse concept, where the terrestrial environment subsidises the aquatic system, as has also been observed in the Okavango Delta (Mosepele et al., 2012). The highest yields in these systems occur not in the main river channels (Hoggarth et al., 1999), but in the ecotone, the dynamic interface between the terrestrial and aquatic systems, making floodplains some of the most productive systems globally (Kolding and Zwieten, 2011). Moreover, understanding the dynamics in fish energetics is critical towards maintaining productive fisheries and sustaining river systems (Arthington, et al., 2005).

Ecological processes in river systems are controlled by the quantity and temporal distribution of stream-flow (Poff, 1992). Subsequently, key factors such as water volume/ depth and velocity regulate ecology and distribution of various fish species (Poff and Alan, 1995). This is the fundamental logic that underpins environmental flows and fish ecology and biology in aquatic systems. The main argument in this chapter is that environmental flows essentially tie the ecology of fish species of Boteti River to its flow regime. According to King et al. (2003), a holistic scenario based approach is the best method to assess the effect of various flows in a river system and use the best possible knowledge to make predictions on ecosystem effects. The first step is to establish knowledge of the biology and ecology (e.g., species composition, diet, etc.) of the fish species in the system (Arthington et al., 2003; Welcomme and Halls, 2004), then integrate this knowledge in a holistic approach to reach consensus on environmental flows (King et al., 2003). This is the procedure that was followed in the development of a trans-boundary diagnostic analysis (socio-economic) of the Okavango River Basin (Barnes, et al., 2009). A similar approach was used in this chapter where the ecology and biology of the Boteti River system fish community was established, the socio-economic value of the fishery determined and then predictions were subsequently made on the potential effect of extreme flooding on the fishery of this river system.

Drought and flood events in floodplain systems have a different effect on various fish species where the so-called white fish (oxygen intolerant species) either migrate to deeper/ permanent water (during droughts) while some black fish (oxygen tolerant species) may also migrate or burrow into the river channel (Hoggarth et al., 1999). The life history strategy of floodplain fish species are attuned to variability in flooding patterns (Lowe-McConnell, 1987) where high flood years generate large fish stocks (Hoggarth et al., 199). Floodplain fish stocks in high flood years are characterised by more than one year class/cohort due to enhanced breeding/spawning conditions (de Graaf et al., 2003). While extreme flooding usually damages infrastructure with a consequent loss in human life (Hoggarth et al., 1999; Poff, 2002), these are generally beneficial to fish production in floodplain and river systems. Furthermore, it has been argued that ecologically, extreme floods "rejuvenate rivers systems by maintaining dynamic ecological structure and function" (Poff, 2002).

Therefore, the main aim of this chapter was to establish baseline information on the ecological and socio-economic aspects of the Boteti River fisheries. This baseline information was then used to predict the potential effects of extreme flooding on the Boteti River system's fish stocks and potential impacts on the riparian community. The prediction of the effect of extreme flooding on fish stocks was done based on Mosepele (2009).

MATERIALS AND METHODS

Study Area

The Boteti River receives an annual inflow of water from the Okavango Delta, which is a small percentage of the total in-flow volume (Gumbricht et al., 2004). Flow into the Boteti River had been intermittent for the past several years, except since 2008 when good floods have kept the river constantly flooded. Generally however, a dry Boteti River would suggest that lakes Xau and the Mopipi reservoir also remained dry, both of which used to sustain substantial fisheries when flooded (Masundire et al., 1998). This capricious nature of the floods in the lower Delta and impermanence of water has resulted in low species diversity (Bruton et al., 1984; Merron and Bruton, 1988; Merron 1991; Nordin and Alonso, 2003) and hence low fisherman densities (Mosepele, 2000). Nonetheless, Skjonsberg and Merafe (1987) revealed that there was a thriving commercial fishery in the Boteti River, when the river had water. Furthermore, Silitshena and McCleod (1989) indicated that Lake Mopipi made up the bulk of fishing enterprises during the dried-salted fish era when income generated was approximately BWP 620.00 fisher^{-1} yr^{-1} around 1986/1987. Notwithstanding, the fishery in this system is vulnerable to changes in the Delta's flooding patterns. When outflow to Thamalakane River from the upper Delta decreased substantially a few decades ago, the Boteti River dried up, and with it so did the fishery that existed at that time (Mosepele, 2003).

Three sites were selected along the Boteti River as illustrated in Figure 1. All these sites were lagoon habitats which would naturally retain water for longer periods than other areas in the river system.

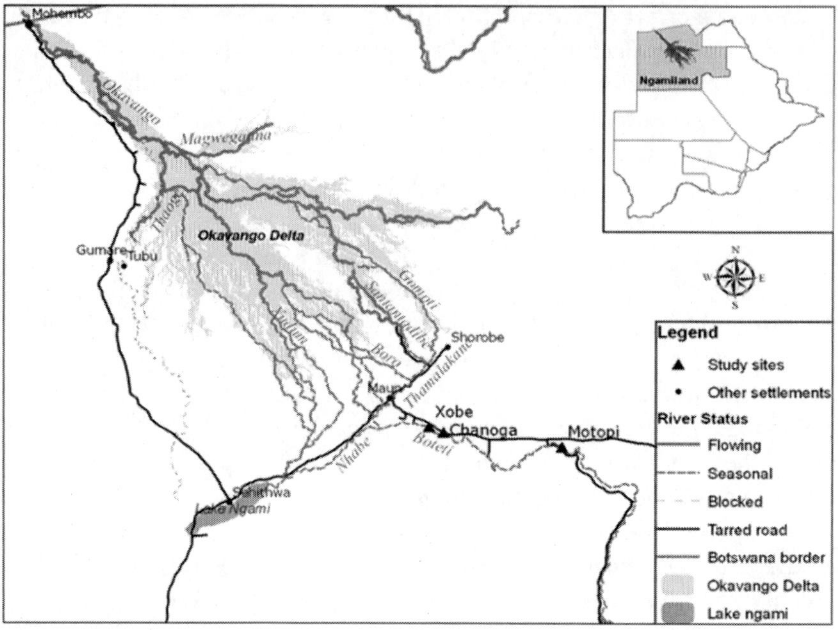

Figure 1. Map of the study area showing the three study sites.

At Xhobe, the lagoon habitat here was characterized by papyrus and phragmites growth which fringed it on both sites. Ludwigia spp was also growing on the edges of the lagoon at this sampling site. The Chanoga lagoon is relatively the most extensive of the sites chosen in this river system, and is also characterized by a profuse growth of papyrus and phragmites on its northern portion. The southern part of the lagoon is characterized by rocky outcrops. Motopi lagoon is a wide expanse of water, fringed on either side by acacia trees. The headwaters of the flooding during this sampling period were just a few kilometres downstream from Motopi. Xhobe is approximately 10 km from Chanoga while Chanoga is approximately 80 km away from Motopi along the river. Therefore, Xhobe is approximately 90 km away from Motopi along the Boteti River.

Data Collection

Fish Species Composition, Abundance and Diversity

Standardised experimental fishing nets were used to sample the fish populations of Boteti River between October 2008 and February 2009 (5 months) at three sites along the river system. Two types of fishing nets, mono-filament and multi-filament were used to sample fish. One type consisted of 12 - three meter long mesh panels, with the mesh sizes arranged in a geometric series: 10, 13, 16, 20, 25, 31, 39, 48, 58, 70, 86, and 110 mm stretched meshes. The multi-filament net is made up of 12, 10-m long panels of mesh sizes; 12, 16, 22, 28, 35,

45, 57, 73, 86, 93, 118, 150 mm stretched meshes. The nets were set in the morning and then collected in the evening (for 10 hours) and set again in the evening and collected in the morning (for 12 hours). After retrieving the nets from the water they were brought back to camp where the fish were carefully removed. Special care was taken to ensure that fish from each panel were kept separate from each other and the mesh size was recorded. Each individual fish specimen was then identified to species, where total length (TL) was subsequently measured, sex and maturity stage for some species determined. Fish species were then sub-sampled for feeding analysis where five individuals of each species in each length class were selected. The entire gut for each specimen was then removed, and preserved in alcohol for lab analysis.

In the laboratory, food items were washed into a petri-dish and prey items identified according to Clegg's (1974) order, separated and counted (in the prey count, fish heads were counted as individuals). The dry weight was determined by drying the prey items at 60°C for 24 hours and weighed in milligrams using a Mettler Toledo AR21 comparator weighing balance.

Total Fishing Effort/ Socio-Economic Survey

A structured questionnaire was conducted in two main commercial fishing camps along the Boteti River to ascertain the demography, ethnicity, and general socio-economic status of fishers in this system. A similar approach was used in the Okavango Delta (Ngwenya and Mosepele, 2008; Mosepele and Ngwenya, 2010) to determine the socio-economic structure and dynamics of the fishing community. Moreover, the survey was also administered to respondents in 9 different villages and informal settlements, including: Samedupi, Tshaadamo, Chanoga, Xhana, Mawana, Xhaega, Segoro, Gudie and Baipidi. Names of known gillnet fishers were obtained from a Fisheries Division database and these were purposively selected for interviews. Furthermore, local people in the study villages were also asked for location of gill net fishers in their areas. In that case, the total count method was used and 20 gillnet fishers were subsequently interviewed.

The structured questionnaire covered all various socio-economic aspects of the fishers; economic status of commercial and subsistence fishers with respect to income generated from fishing and how the income is utilized; complementarity of fishing to other household livelihood strategies, or competition for resources especially with agriculture for time and labour; division of labour amongst family members, particularly women and children for various processes of a fishery activity such as fishing, marketing and equipment repair; efficiency of current fish marketing infrastructure and value of bartering of fish for other products, especially grain; value of fishing as a food security measure more especially in times of crop failure and livestock disease outbreaks.

Total Yield from the Fishery

This was determined from daily catch and effort data that gill net fishers in the river system record daily and they are collected monthly by the Fisheries Division in the Department of Wildlife and National Parks from the Ministry of Environment, Wildlife and Tourism. The data were then captured in Pasgear (Kolding and Asmund, 2010) software package.

Data Analysis

Fish Species Composition and Relative Importance

A measure of the relative abundance or commonness of each species (i) in the catch composition was calculated using an index of relative importance, IRI, (Kolding, 1989):

$$\%\,\mathrm{IRI_i} = \frac{(\%\mathrm{W_i} + \%\,\mathrm{N_i}) \cdot \%F_i}{\sum_{j=1}^{S} (\%W_j + \%\mathrm{N_j}) \cdot \%F_j} \cdot 100$$

where, % W_i and % N_i is percentage weight and number of each species of total catch; % F_i is percentage frequency of occurrence of each species in total number of settings; and S is total number of species

Shannon's Diversity Index (H') and relative evenness (J'): These are defined in Kolding (2000) as:

$$H' = -\sum_{i=1}^{S} P_i \cdot Ln(P_i)$$

where is the relative abundance of individuals found in the i- th species.

$$J' = H' / H_{\max}$$

where

$$H' = -\sum_{i=1}^{S} P_i \cdot Ln(P_i)$$

Catch Rates and Biomass

$$\mathrm{CPUE} = \frac{1}{\mathrm{y}} \sum_{\mathrm{i=1}}^{n} w_{\mathrm{i}} \cdot \frac{\mathrm{SU}}{\mathrm{U_i}} \cdot \frac{\mathrm{ST}}{\mathrm{T_i}}$$

where, y = effort defined as number of settings (nets); n= number of samples (observations); w_i= catch (in weight or numbers) in set_i or $sample_i$; SU = standard relative effort unit (size) of a net panel; U_i = actual effort unit (size) of net_i; ST = standard time unit of a setting; T_i = actual time unit of a setting.

Total Fishing Effort/ Socio-Economic Data

Various frequencies to establish the total fishing effort and various socio-economic variables were calculated using the SPSS computer software program from questionnaires that were administered to respondents in the study area.

Other Tests

Furthermore, single factor Anova was used to explore the level of significance between various biological variables in this study. The Spearman rank order correlation was also used to test the strength and significance level of relationships between observed biological variables during this study.

Extreme Flooding Prediction

Environmental flow assessment is one of the best tools to predict/ simulate effects of variable flows on fish communities (Arthington et al., 2003; King et al., 2003). The only environmental flows assessment for fish populations in Botswana was conducted by Mosepele (2009). Six indicators were developed (based on guilds) for the Okavango Delta and the Boteti River system. However, only two indicators were adopted (see Table 1 below) and used for the Boteti River system because it does not have extensive habitats (e.g., sandbanks, floodplains and others) like the rest of the Delta.

Table 1. Summary of environmental flow indicators developed for the Boteti River

Indicator	Description	Representative Species
1	Resident channel dwellers	*S. robustus, S. giardi, M. macrolepidotus, S. intermedius, T. rendalli, Oreochromis spp, Serranochromis spp, C. gariepinus*
2	Marginal vegetation dwellers	*P. castelnaui, S. nigromaculatus, S. carlottae, S. macrocephalus, H. odoe, B. lateralis, B. poechii, Aplocheilichthys spp., B. afrovernayi, B. multilineatus, P. philander, H. elongatus, T. sparrmanii, P. acuticeps, B. paludinosus, M. intermedium*

Source: Adapted from Mosepele, (2009).

Response curves were then developed for these fish species/ guilds using an environmental flows decision support system (DSS) developed for the Okavango River Basin. This procedure is described extensively in King et al., (2009) and was first proposed and developed by King et al., (2003). The procedure for developing response curves for the fish component of this approach is described in Arthington et al., (2003).

RESULTS

Catch Composition, Relative Importance and Diversity

The five month experimental fishing survey in the Boteti River sampled 39 different fish species among the three stations (Table 2).

Table 2. Fish species catch composition and diversity from three sampling stations along the Boteti River

Species	Xhobe			Chanoga			Motopi			Overall	
	IRI	H`	J`	IRI	H`	J`	IRI	H`	J`	IRI	FO
Brycinus lateralis	3568	0.339		3246	0.367		4944	0.303		3796	53.7
Barbus poechii	199	0.146		2799	0.346		262	0.172		954	33.7
Marcusenius macrolepidotus	781	0.22		432	0.129		9	0.053		334	19.1
Tilapia sparrmanii	624	0.202		199	0.113		114	0.095		287	29.8
Barbus unitaeniatus	12	0.057		557	0.272		210	0.217		212	24.3
Schilbe intermedius	747	0.175		101	0.055			-		179	20.9
Barbus paludinosus	62	0.102		248	0.184		111	0.13		141	20.7
Sargochromis codringtonii	94	0.105		24	0.043		239	0.142		93	15.2
Clarias gariepinus	85	0.027		3	0.004		324	0.025		84	5.7
Serranochromis macrocephalus	48	0.054		39	0.034		44	0.068		43	10.7
Oreochromis andersonii	12	0.017		3	0.007		204	0.094		34	7.4
Serranochromis thumbergi	14	0.03		26	0.031		60	0.061		28	12.2
Pharyngochromis acuticeps	1	0.011		21	0.047		65	0.107		19	6.3
Clarias ngamensis	10	0.013		10	0.01		14	0.013		11	3.9
Petrocephalus catastoma	74	0.114		0	0.006			-		10	7.2
Oreochromis macrochir	21	0.03		3	0.008		12	0.025		10	6.5
Serranochromis angusticeps	11	0.02		2	0.007		4	0.012		5	4.8
Tilapia rendalli	10	0.024		3	0.012		1	0.01		4	4.6
Sargochromis greenwoodii	6	0.029		4	0.02		1	0.016		4	3.9
Sargochromis carlottae	13	0.032		1	0.009		0	0.005		3	3.9

Species	Xhobe			Chanoga			Motopi			Overall	
Hepsetus odoe	6	0.013		1	0.003			-		2	1.1
Pseudocrenilabrus philander	11	0.041			-		0	0.003		1	4.1
Serranochromis altus	1	0.006		2	0.007		0	0.012		1	1.7
Synodontis leopardinus	1	0.011		1	0.007		0	0.005		1	2.6
Barbus barnardi	1	0.024		2	0.035			-		1	2
Clarias theodorae	4	0.015		0	0.004			-		1	2
Serranochromis robustus	0	0.004		0	0.006		0	0.003		0	1.1
Tilapia ruweti	2	0.021			-			-		0	0.7
Squeakers (synodontis sp)		-			-		2	0.034		0	0.4
Synodontis macrostigma		-		1	0.012			-		0	0.4
Sargochromis giardi	1	0.011			-			-		0	0.7
Barbus thamalakanensis	0	0.002		0	0.012			-		0	0.9
Synodontis woosnami	0	0.006			-			-		0	0.7
Synodontis macrostoma	0	0.002		0	0.003			-		0	0.7
Synodontis nigromaculatus		-		0	0.003			-		0	0.4
Synodontis thamalakanensis	0	0.006			-			-		0	0.7
Hippopotamyrus discorhynchus	0	0.014			-			-		0	0.2
Synodontis vanderwaali		-		0	0.002			-		0	0.2
Barbus bifrenatus	0	0.004			-			-		0	0.2
Hemigrammocharax multifasciatus	0	0.004			-			-		0	0.2
Total		1.931	0.54		1.799	0.52		1.603	0.51		

Table 3. Summary of various statistical tests exploring the level of significance of various indices of selected species among the three sampling stations where *p* values in bold indicate a significant difference at the 95% confidence interval

	Index	DF	F	P	Xobe	Chanoga	Motopi
C. gariepinus	no/set	12	3.28	0.1	0.91		0.42
	g/set	8	0.15	0.71	2244.03		2863.6
	ML	13	0.79	0.39	499.28		602.78
S. intermedius	no/set	33	5.71	**0.02**	6.07	1.43	
	g/set	32	8.36	**0.01**	428.92	105.94	
	ML	36	5.89	**0.02**	213.54	236.59	
M. macrolepidotus	no/set	16	7.79	**0.01**	5.28	2.06	
	g/set	14	0.002	0.97	213.23	211.58	
	ML	25	7.23	**0.01**	180.94	207.2	
B. lateralis	no/set	18	3.64	**0.05**	23.14	13.54	23.61
	g/set	12	6.13	**0.02**	308.28	223.52	184.86
	ML	51	0.64	0.53	107.57	113.49	108.99
O. andersonii	no/set	17	0.36	0.7	0.70	0.44	0.64
	g/set	16	7.37	**0.01**	262.57	49.17	122.64
	ML	24	5.06	**0.02**	262.43	225.78	177.43
All species	Diversity	89	0.98	0.38	1.93	1.80	1.60
	no/set	50	1.08	0.35	20.03	30.01	26.88
	g/set	53	2.16	0.13	678.94	733.21	852.03
	ML	69	2.37	0.10	126.15	120.26	118.45

Based on Shannon's diversity index (H'), Xhobe had the most diverse fish community followed consecutively by the other stations. A characin, *Brycinus lateralis* was the most important species in all the three sampling stations. The second most important species in Xhobe was a momyrid, *Marcusenius macrolepidotus*, while a cyprinid, *Barbus poechii* was the second most important in Chanoga and a catfish, *Clarias gariepinus* was second most important in Motopi. Furthermore, there is a progressive decrease in fish species diversity (and evenness) from Xhobe to Motopi sampling stations.

As summarized in Table 3, there was no significant difference at the 95% confidence interval (i.e., p>0.05) in grams/ set, diversity, no/set or mean length between the three sampling stations in the study area. This suggests that there was homogeneity and fluidity in the overall fish community of this river system within the study time frame. Furthermore, there was no habitat partitioning by mean length of selected species in the study area. However, some significant differences were observed within the fish community at the species level among the three habitats.

However, there were significant differences (p<0.05) in all indices for *S. intermedius* between Xhobe and Chanoga where the density (p=0.02) and abundance (p=0.01) were highest in Xhobe while the mean length (p=0.02) was biggest in Chanoga. Similarly, mean density (p= 0.01) for *M. macrolepidotus* was significantly higher at Xhobe than Chanoga while mean length (p=0.01) was significantly higher at Chanoga than Xhobe. Mean abundance (p=0.01) and mean length (p=0.02) for *O. andersonii* were significantly highest at Xhobe among the three habitats while there were no significant differences in mean density among the three habitats.

Table 4. Species checklist for Boteti River

Species	1	2	3	4	5	6
Aplocheilichthys johnstoni			x			x
Barbus afrovernayi			x			X
Barbus barnardi			x	x	x	X
Barbus bifrenatus			x		x	X
Barbus fasciolatus			x			X
Barbus multilineatus			x			X
Barbus paludinosus			x	x	x	X
Barbus poechii	x	x	x	x	x	X
Barbus radiatus			x			X
Barbus thamalakanensis			x	x	x	X
Barbus unitaeniatus			x	x	x	X
Brycinus lateralis	x	x	x	x	x	X
Clarias gariepinus	x		x	x	x	X
Clarias ngamensis		x	x	x	x	X
Clarias theodorae			x	x	x	X
Hemichromis elongatus			x			X
Hemigrammocharax machadoi			x			X
Hemigrammocharax multifasciatus			x		x	X
Hepsetus odoe	x	x	x	x	x	X
Hippopotamyrus discorhynchus			x		x	X
Marcusenius macrolepidotus	x	x	x	x	x	X
Momyrus lacerda	x	x	x			X
Oreochromis andersonii	x		x	x	x	X
Oreochromis macrochir	x	x	x	x	x	X
Parauchenoglanis ngamensis			x			X
Petrocephalus catastoma		x	x	x	x	X
Pharyngochromis acuticeps			x	x	x	X
Pollimyrus castelnaui			x			X
Pseudocrenilabrus philander			x		x	X
Rhabdalestes maunensis			x			X
Sargochromis carlottae		x	x	x	x	X
Sargochromis codringtonii		x	x	x	x	X
Sargochromis giardi			x		x	X
Sargochromis greenwoodii			x	x	x	X
Schilbe intermedius	x	x	x	x	x	X
Serranochromis altus				x	x	X
Serranochromis angusticeps		x	x	x	x	X
Serranochromis longimanus		x	x			X
Serranochromis macrocephalus	x	x	x	x	x	X
Serranochromis robustus			x	x	x	X
Serranochromis thumbergi			x	x	x	X
Synodontis leopardinus	x	x	x	x	x	X
Synodontis macrostigma		x	x	x	x	X
Synodontis macrostoma				x	x	X

Table 4. (Continued)

Species	1	2	3	4	5	6
Synodontis nigromaculatus	x	x	x	x	x	X
Synodontis thamalakanensis					x	X
Synodontis vanderwaali				x	x	X
Synodontis woosnami		x	x		x	X
Tilapia rendalli			x	x	x	X
Tilapia ruweti			x		x	X
Tilapia sparrmanii			x	x	x	X
Species Richness	12	18	47	31	39	51

Note: 1 = Merron et al. 1984; 2 = Merron and Bruton, 1984; 3 = Merron and Bruton, 1988; 4 = This study (Chanoga only); 5 = This study (Boteti River); 6 = Species checklist for Boteti.

A species checklist of the Boteti River is summarized in Table 4 using data from this study and previous studies in this system. There was relatively low species richness in the 1984 surveys compared to the 1983-1986 study and the present study. These differences in richness are attributed to differences in sampling frequency in the river system and possibly different sampling gears also. Notwithstanding, the table suggests that there are potentially 51 different fish species in the Boteti River, though occurring at different abundances, where some species such as *B. lateralis* are dominant in the system, while other species such as *S. longimanus* are rare. Spatio-temporal variations: There were spatial variations in catch rates, mean length, diversity and biomass at the three sampling stations in the Boteti River (Figure 2).

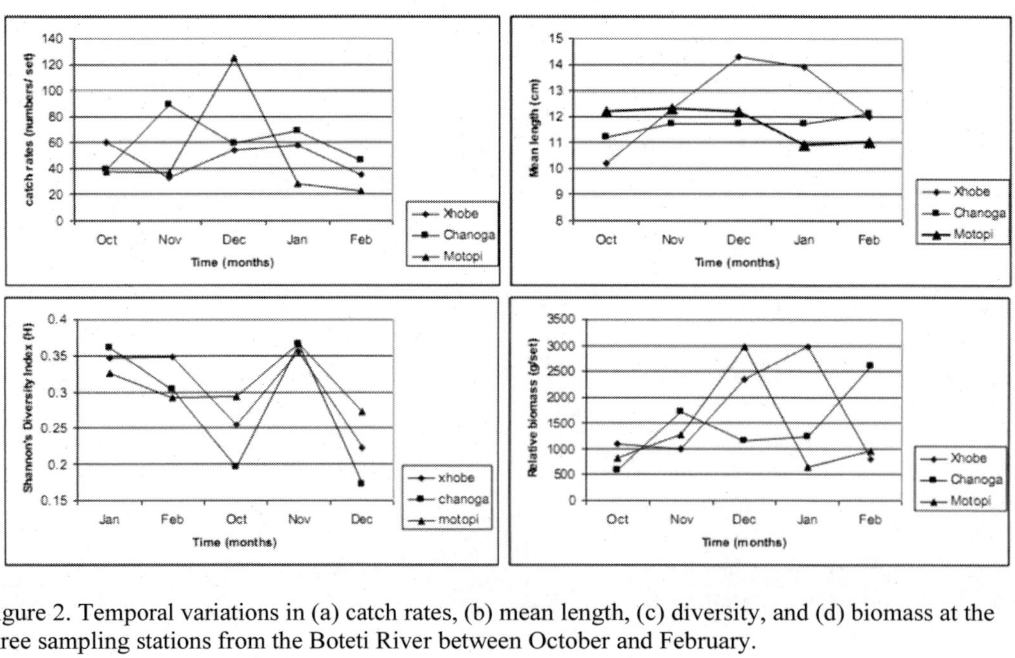

Figure 2. Temporal variations in (a) catch rates, (b) mean length, (c) diversity, and (d) biomass at the three sampling stations from the Boteti River between October and February.

Catch rates were highest in Xhobe in October, followed by Chanoga in November and Motopi in December, whereupon they declined in all stations towards February. These dynamic relationships are further explored in Table 5 through spearman rank order correlations. The most dynamic relationships occur between Xhobe and Motopi sites for catch rates and relative biomass. While the relationships are not significant (p>0.05), there is a high negative relationship between catch rates and relative biomass between the two sites. Significant correlations (p=0.04) were observed in species diversity between Xhobe vs. Chanoga and between Chanoga vs. Motopi while no significant correlations were observed between Xhobe and Motopi (Table 5).

Table 5. Spearman rank order correlations, with *p* value where bold indicate a significant relationship

Test	no/set	length	diversity	biomass
Xobe *vs*. Chanoga	-0.6(0.28)	0.22(0.72)	0.9 (0.04)	-0.6 (0.28)
Xobe *vs*. Motopi	0.3(0.62)	-0.1(0.87)	0.7(0.19)	-0.3(0.62)
Chanoga *vs*. Motopi	-0.1(0.87)	-0.34(0.57)	0.9(0.04)	0.1(0.87)

Feeding Ecology (Selected Species)

The three most important food items in the Boteti River during the study period were chironomids, fish and algae (Table 6), respectively.

Table 6. Overall % IRI of food items for several fish species from Boteti River

Food item	%IRI
Chironomidae	44.7
Fish	37.3
Algae	11.7
Libellulidae	5.2
Libellulidae and algae	0.3
Detritus	0.3
Unknown insects	0.2
Tabanidae	0.1
Chironomidae and Libellulidae	0.1
Terrestrial insects	0.1
Grasshopper	<0.1
Water lily seeds	<0.1
Snail	<0.1
Dyticidae larvae	<0.1
Grass	<0.1
Sedge	<0.1
Seeds	<0.1
Vegetation	<0.1
Libellulidae and grass	<0.1
Dragonfly	<0.1
Fish and algae	<0.1
Bird feathers	<0.1
Libellulidae and seeds	<0.1
Propistomatidae	<0.1
Aeshinidae	<0.1
Psychodidae	<0.1
Libellulidae and tabanidae	<0.1
Pleidae	<0.1
Psephenidae	<0.1

Table 6. (Continued)

Food item	%IRI
Terrestrial insects and algae	<0.1
Ceratopogonidae	<0.1
Libellulidae and feathers	<0.1
Algae and unknown insects	<0.1
Mosquito larvae	<0.1
Leptoceridae	<0.1
Feathers	<0.1

Out of the approximately 36 food groups identified, chironomids were the most dominant with the highest IRI value (≈45%), while the IRI value of 26 food groups was less than 1%. Nonetheless, there was a wide variety of diet items ranging from seeds to bird feathers and terrestrial insects. This suggests that there is a wide variety of food items available to fish in this system from both aquatic and terrestrial sources.

Table 7. Temporal variations in the % Index of Relative Importance (IRI) of food items for *B. lateralis* where values in bold are the dominant food items

Species	Oct	Nov	Dec	Jan	Feb
Libellulidae	56.4	79.3	5.9	88	89.5
Algae	13.2	4.4	59.6		
Unknown insects	3	1.3	25.2	1.4	
Libellulidae and algae	21.8	0.5	1.2		
Dysticidae larvae		11.4			
Libellulidae and seeds	2.7			1.8	
Fish			8.1		
Propistomatidae		1.3			
Libellulidae and tabanidae		1			
Grass	1.1				
Terrestrial insects and algae				2.6	
Bird feathers				2.5	
Libellulidae and feathers				2.3	
Algae and unknown insects	0.7				
Dragonfly		0.4			
Terrestrial insects				1.3	
Chironomidae					10.5
Leptoceridae		0.3			
Sedge	0.5				
Feathers	0.4				
Seeds	0.1				

B. lateralis: Libellulidae was the most important food item for *B. lateralis*, except in December where algae was dominant in its diet (Table 7). Nonetheless, *B. lateralis* had a very diverse diet towards (Figure 4) the end of the year (October and November) but its food items became gradually less diverse over time, until February where Chironomids and Libellulidae were the only prey items. There is a gradual shift in food sources from a mainly aquatic food base between October and December to terrestrial prey items in January. Furthermore, feeding rates were generally higher between October and November (with lower frequencies of empty stomachs), were lowest between December and January, before increasing again (Figure 3). The observed temporal variations in feeding rates and diversity of food items were, however, not significant at the 95% confidence level (Table 7).

S. intermedius: Fish was the most important food items for *S. intermedius*, its dominance increased gradually over time and was highest in February (Table 8). This gradual shift in diet also corresponded to a decrease in the contribution of aquatic invertebrates to its diet. This change in diet was more pronounced in February when terrestrial insects began to appear in its diet. Interestingly, in December when the frequency of empty stomachs was highest (Figure 5a), *S. intermedius* diet consisted of a relatively high proportion of a low energy food items, algae, and some atypical food items such as snails and seeds.

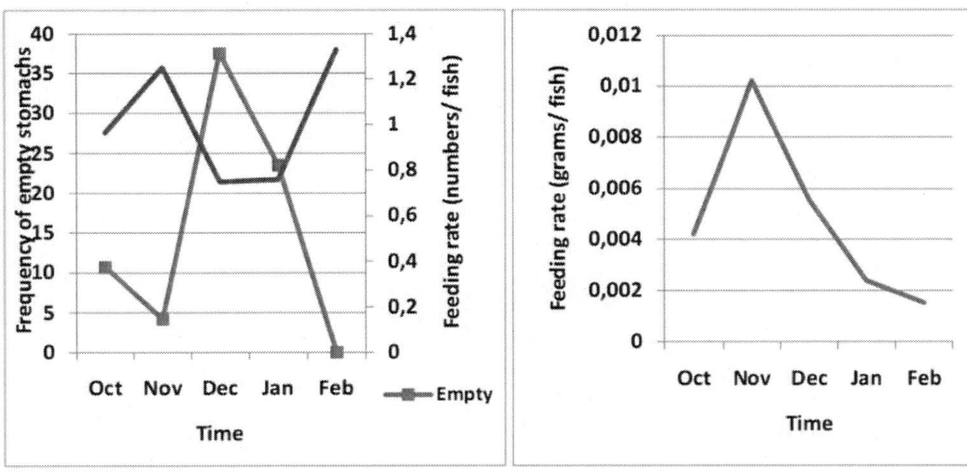

Figure 3. (a) Temporal variations in the feeding rates (numbers/ fish) behaviour of *B. lateralis* and frequency of occurrence of empty stomachs, (b) Temporal variations in the feeding rate of *B. lateralis* using weight of food items/ fish.

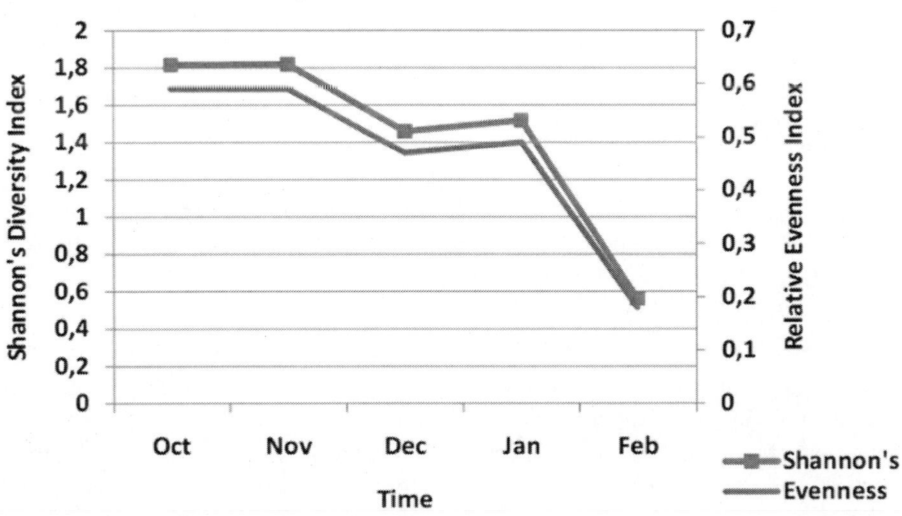

Figure 4. Temporal variations in the diversity of food items for *B. lateralis*.

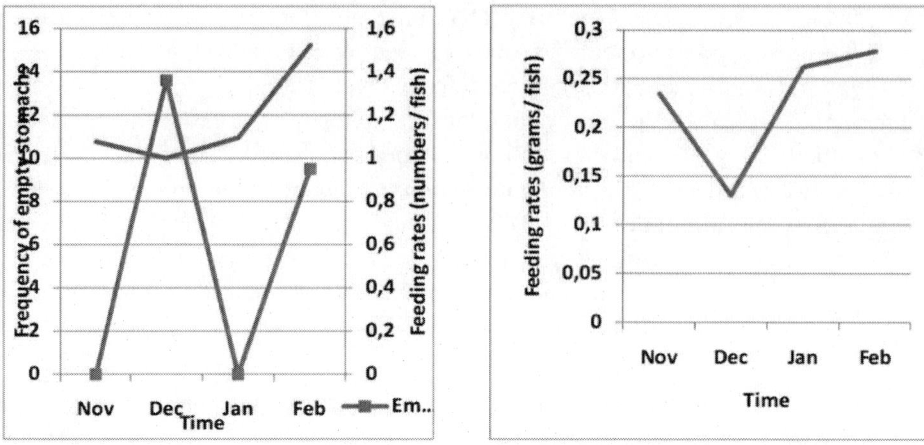

Figure 5. (a) Temporal variations in the feeding rates (numbers/ fish) behaviour of *S. intermedius* and frequency of occurrence of empty stomachs over time, (b) Temporal variations in the feeding rate of *S. intermedius* using weight of food items/ fish.

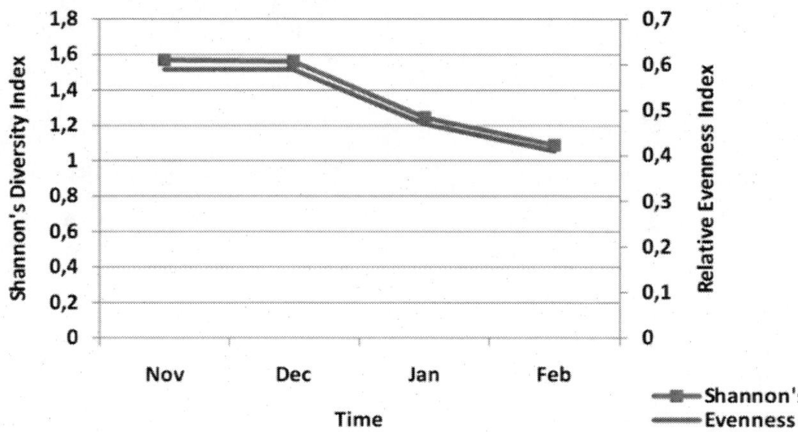

Figure 6. Temporal variations in the diversity of food items for *S. intermedius*.

These changes in diet composition also contributed to a gradual decrease in the diversity of food items (Figure 6), even though this decrease was not significant (p = 0.372). The feeding rates (grams/ fish) of *S. intermedius* were lowest in December, when the relative importance of fish in its diet was not very high compared to other months. Similarly, feeding rates (grams/ fish) were highest in February when fish contributed much more to its diet than in the other months (Figure 5b).

M. macrolepidotus: Chironomids were the major diet items for *M. macrolepidotus* throughout the study period, and algae also constituted an important part of its diet (Table 9). While the rate of prey items eaten by M. *macrolepidotus* increased over time from December (though not significant at p= 0.255), the highest frequency of empty stomachs was observed in January (Figure 8a). Interestingly, this period coincides with a complete dominance of the diet by Chironomids, where they comprised over 94 % of IRI. However, feeding rates expressed in grams/ fish declined sharply (but were not significant at 95% confidence level) between November and December before they increased slightly until February (Figure 8b).

Generally over the entire study period, there was a significant (p=0.000) decline in the diversity of food items eaten by *M. macrolepidotus*.

Table 8. Temporal variations in the % Index of Relative Importance (IRI) of food items for *S. intermedius* where values in bold are the dominant food items

Prey item	Nov	Dec	Jan	Feb
Fish	68.2	66.4	88.8	89.3
Libellulidae	22.1	10.9	10.2	8.5
Algae	5	14.6		0.2
Grasshopper	2.4	0.7	0.3	0.3
Unknown insects	1.1	3.4		
Terrestrial insects				1.5
Snail		3.8		
Fish and algae	1.3		0.3	
Libellulidae and algae				0.3
Pleidae			0.2	
Chironomidae			0.2	
Seeds		0.2		

Table 9. Temporal variations in the % Index of Relative Importance (IRI) of food items for *M. macrolepidotus* where values in bold are the dominant food items

Species	Nov	Dec	Jan	Feb
Chironomidae	30.8	40	94.9	47.6
Algae	7.7	25	2.5	33.3
Libellulidae	7.7	15	1.2	14.3
Chironomidae and Libellulidae	7.7			
Libellulidae and algae	15.4		<0.1	4.8
Tabanidae		5	1	
Vegetation		5	0.1	
Grass			0.2	
Seeds	7.7			
Libellulidae and grass			0.1	
Unknown insects			<0.1	
Aeshinidae	7.7			
Pshychodidae	7.7			
Fish		5		
Ceratopogonidae		5		
Mosquito larvae	7.7			
Snail			<0.1	

Other key species: There were temporal shifts in the *O. andersonii* diet where it switched its diet from a predominantly high proportion of detritus in November, to an algae dominated diet in January. Furthermore, algae and detritus were the only two items observed in the diet of *O. andersonii* in this system. Furthermore, Figure 4 illustrates that while *C. gariepinus* had a relatively diverse diet, fish were the dominant food item for this species in the Boteti River system.

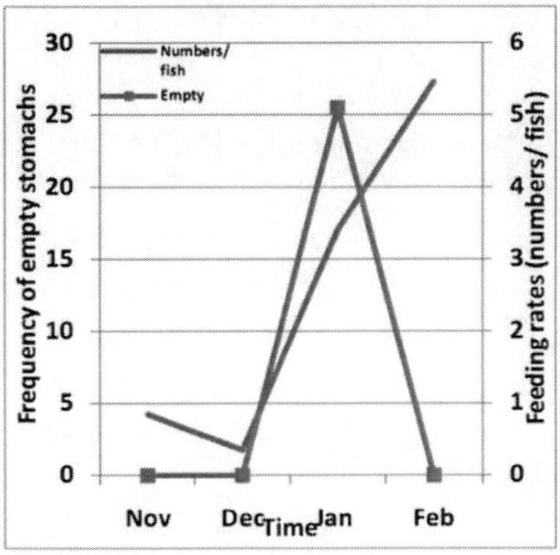

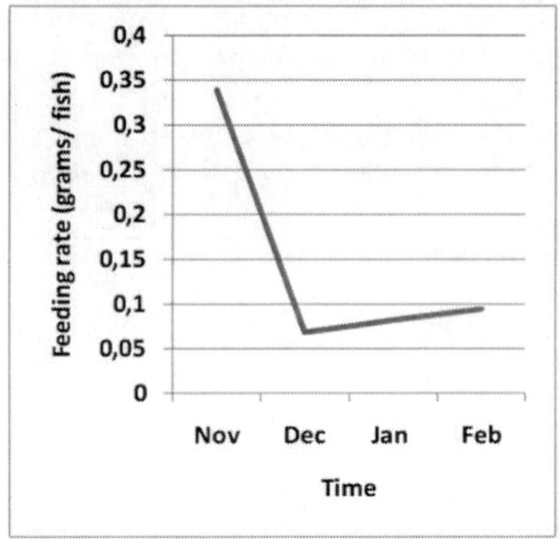

Figure 7. (a) Temporal variations in the feeding rates (numbers/ fish) behaviour of *M. macrolepidotus* and frequency of occurrence of empty stomachs over time, (b) Temporal variations in the feeding rate of *M. macrolepidotus* using weight of food items/ fish.

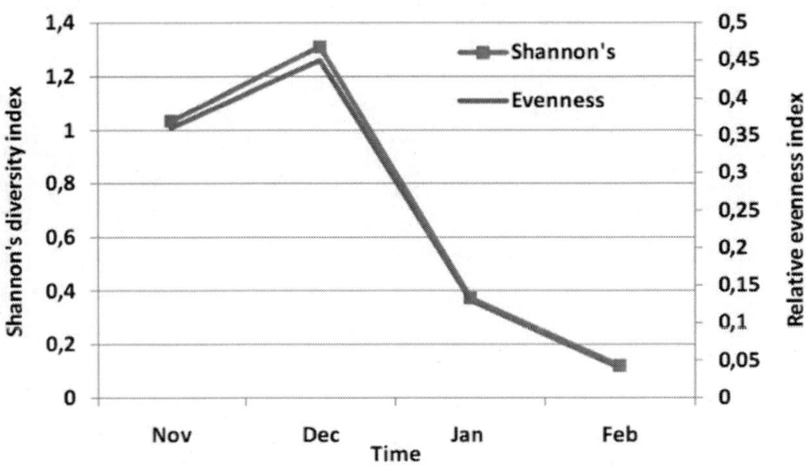

Figure 8. Temporal variations in the diversity of food items for M. macrolepidotus.

Table 10. Summary of Anova test to explore the level of significance on feeding rates (numbers/ fish and weight/ fish) and diet diversity among different months of the study period (i.e., October – February) where *p* values with an * indicate significance difference at 95% confidence level

Species	Diversity	Numbers/ fish	Weight/ fish
B. lateralis	0.304	0.884	0.396
C. gariepinus	0.930	0.613	0.676
S. intermedius	0.372	0.919	0.926
M. macrolepidotus	0.000*	0.255	0.587

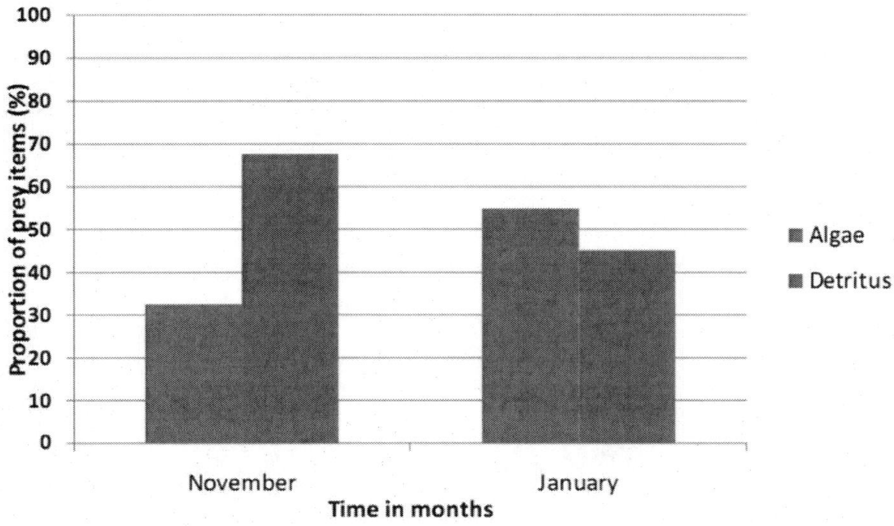

Figure 9. Proportion of food items of *O. andersonii* from Boteti River.

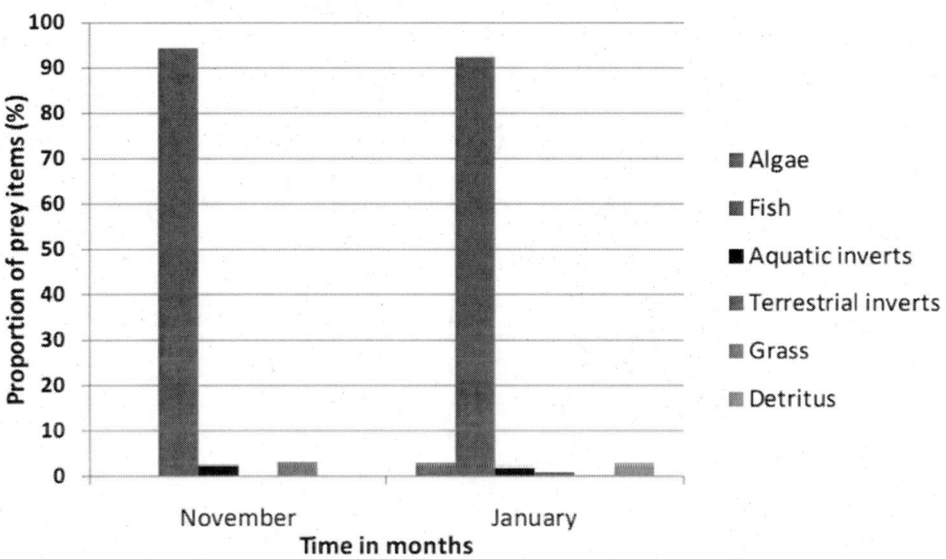

Figure 10. Proportion of food items of *C. gariepinus* from Boteti River.

Total yield from the gill net fishery: Breams/ tilapia group is the most important (based on the Index of Relative Importance - IRI) in the catch composition of Boteti Rivers fishers (Table 11).

This group constitutes just below 50% by total weight of all the catches while *H. vittatus* is the least important species in the gill net catches. The catfish group is the second most important group followed by an others group (this group includes all the species that occur in the system and can be harvested within the mesh size range in this fishery: 100 – 125 mm stretched mesh size range) Bream catch rates (no/set) normally follow a temporal pattern where relatively high rates are observed in March and August while the lowest catch rates are observed in October (Figure 11).

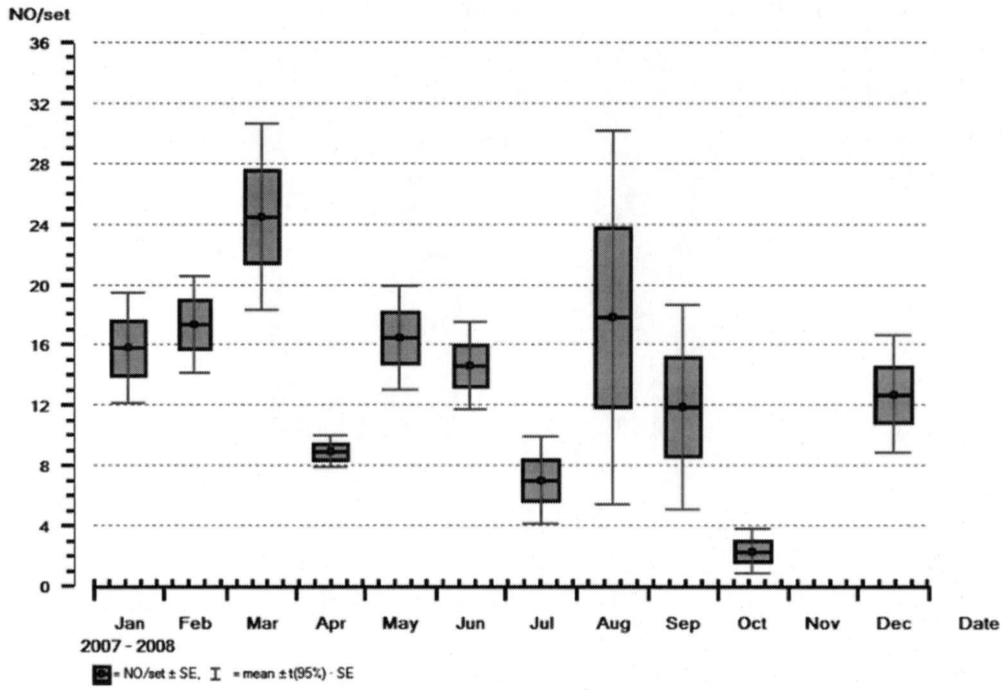

Figure 11. Temporal variations in catch rates (no/ set) of breams from gill net fishers from the Boteti River between 2007 and 2008.

Table 11. Summary of catch composition of gill net catches from Boteti River fishers between 2007 and 2008

Species	NO	% NO	W(tons)	% W	FRQ	% FRQ	IRI	% IRI
Breams/ Tilapia[1]	52696	65.7	22.659	49.7	1164	90.1	10396	62.9
Catfishes[2]	22006	27.5	21.995	48.2	1016	78.6	5949	36
Others[3]	3389	4.2	0.56	1.2	309	23.9	130	0.8
S. intermedius	2056	2.6	0.41	0.9	183	14.2	49	0.3
Hydrocynus vittatus	11	0	0.013	0	9	0.7	0	0
Total	80160	100	45.637	100	-	-	16524	100

N.B: 1: This group includes all the cichlid species that can be harvested within the mesh size range used in this fishery. 2: This group is normally dominated by two Clarias species, *C. Gariepinus* and *C. ngamensis*. 3: This group includes but is not limited to, Synodontis group, Momyrids, etc.

Extreme flooding prediction: Based on Table 12, short term effects of extreme flooding on indicators 1 and 2 during transition 1 indicate that fish larvae and eggs will most probably be washed away by extreme flooding. However, higher floods than normal, associated with a shallow hydrograph during transition 2 phase are generally beneficial to fish populations in indicator/ guild 2. Conversely, extreme floods during the flood season generally have a negative effect on fish populations of indicator 2 as summarized in Table 12.

Table 12. Predicted response to changes fish indicators as a response to extreme flooding

Indicators	Season	Flood prediction	Predicted response of indicator
Indicator 1			
	Transition 1	Steep hydrograph (i.e., extreme flood)	Negative effects. May flush out fish eggs and larvae
			A longer rise will have positive effects
	Flood season	Higher floods than normal	Generally good for fish production.
	Transition 2	Shallow hydrograph	Beneficial to fish larval growth and recruitment
Indicator 2			
	Transition 1	Steep hydrograph (i.e., extreme flood)	Negative effects. May flush out fish eggs and larvae
	Flood season	Higher floods than normal	Negative effects because it can affect fish habitat integrity.
	Transition 2	Shallow hydrograph	A shallow hydrograph might be beneficial to larval fish growth and recruitment

Source: Adapted from Mosepele (2009). Transition 1= increasing flood phase of a hydrograph; Transition 2 = flood decreasing phase of a hydrograph.

Socio-Economic Survey

Demographic Information

Commercial fishers in the Boteti River system are distributed among six different ethnic groups (Figure 12). The BaYei are the dominant fisher group in the area.The average household size is around 5.4 persons, while households are relatively young with 55% of households having 0-5 year-olds; 55% have 5-10 year-olds, and 50% have 11-19 year-olds. Only 25% of households contain people aged 55 and up. Only one fisherman is older than 55 years, and he had been fishing in the 1970's before the river dried. All the other fishermen had only established their fishing businesses within the last one or two years (of the study period).

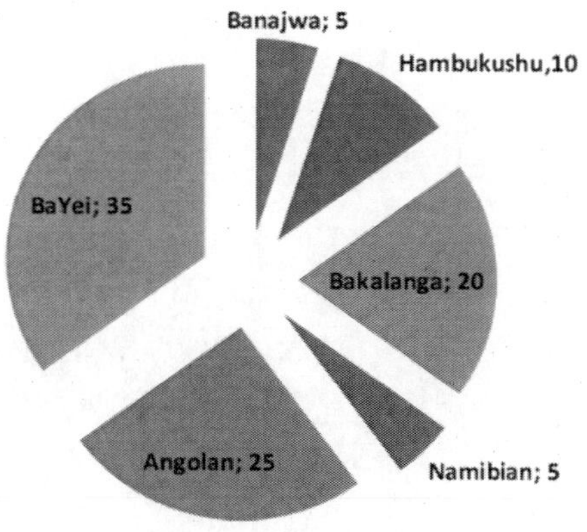

Figure 12. Ethnic Distribution Among Boteti River Commercial Fishermen (%).

Household and Fishing Assets

The majority of fishermen (65%) have no ablution facilities while the rest use pit latrines. Most of the households use the river channel as their main water source, with 35% using a communal tap. However, a small proportion of fishers (10%) indicated they had a water tap in their yard while only 5% of households had a water tap inside the house. Firewood is the commonest source of energy, except for one fisherman who uses a gas stove (but only during the rainy season). The main light source is paraffin lamps (55%), while 35% households also use candles, with a small proportion using wood and battery-powered flashlights (figure 13).

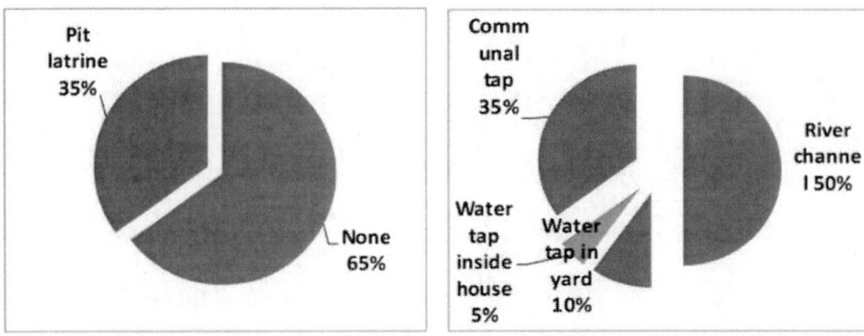

Figure 13. Continued on next page.

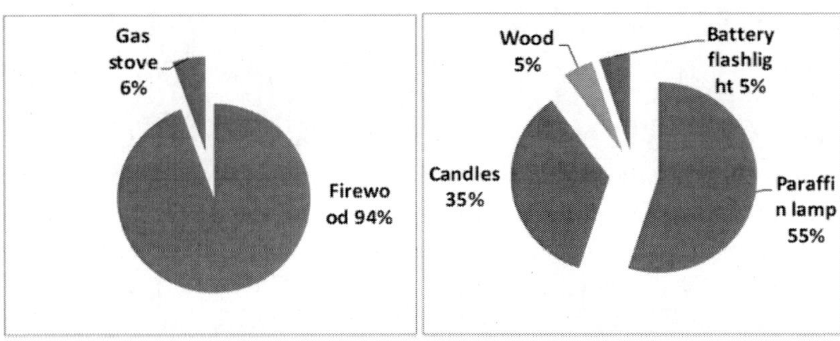

Figure 13. Proportion of household (a) ablution facilities, (b) water source, (c) energy source, and (d) light source.

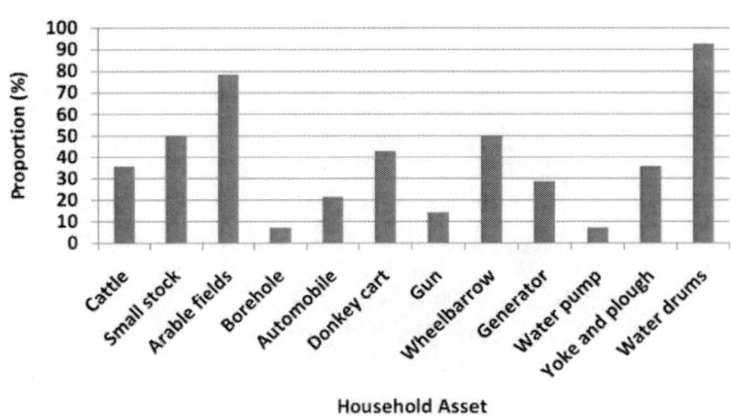

Figure 14. Proportion of household assets' ownership.

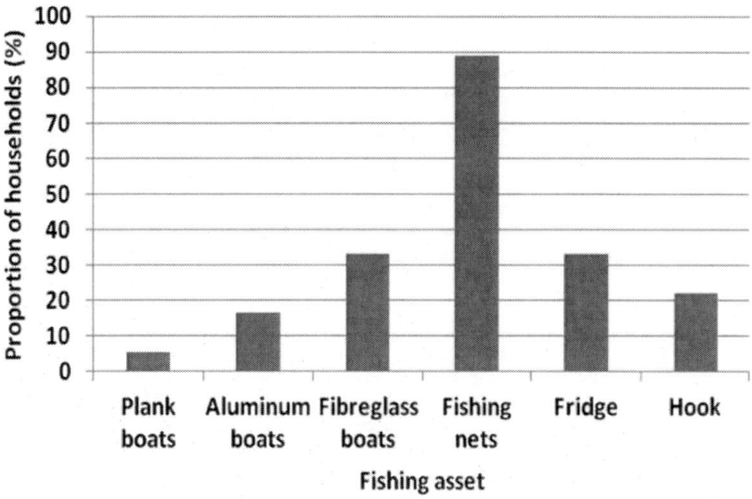

Figure 15. Proportion of households by fishing assets owned.

The most commonly owned household assets are arable fields (over 70%) and (steel) water drums (over 90%). A smaller proportion of fishers own high valued agricultural assets like cattle (just over 30%), while a slightly higher proportion of fishers (just below 50%) own small stock. Moreover, a much lower proportion of fishers own very high valued assets such as automobiles, water pumps and guns (Figure 14).

Fibreglass boats were the commonest fishing craft used by fishers in the river system while wooden boats were the least common. Furthermore, gill nets were the commonest fishing gear owned by fishers while a small proportion (about 20%) of fishers owned fishing hooks also. Moreover, just over 30% of fishers also owned gas powered refrigerators (Figure 15).

Sources of Income

Apart from fishing, fishermen have a diverse basket of income generating activities (Table 13) and it is interesting to note that two of the commercial fishermen have formal sector employment. That notwithstanding, a relatively high proportion of fishermen also have temporary employment and social welfare benefits in addition to fishing (Table 13). Since some of the fishers are refugees from the Dukwe refugee camp, they receive a food basket once a month which qualifies as social welfare.

Fishing has the highest average income per activity among all the other livelihood activities that fishers are involved in (Figure 16). It earns an average of over BWP2500.00/ month (US$376.51), over BWP700.00 (US$105.42) more than formal sector jobs, which is the next highest income earning activity (Note: US$1 = BWP6.64 using 2008 rates). Of course, with commercial fishermen it is assumed that they fish at least 5 months out of the calendar year and it therefore probably contributes a large amount to annual income as well. The money gained from these activities is spent in several ways throughout the month. The greatest monthly expenditure is on food, but transportation and education are other common expenditures (Table 14).

Table 13. Summary of income generating activities that Boteti River fishers are involved in as part of their livelihood activities

Income generating Activities		*Proportion (%)*
Formal Sector Wages	*2*	*10*
Temporary/Piece Jobs	*6*	*30*
Street Vendor	*1*	*5*
Carpentry	*1*	*5*
Thatching	*2*	*10*
Brick making	*1*	*5*
Bakery	*1*	*5*
Poultry Production	*2*	*10*
Fishing	*19*	*100*
Crop Sales	*3*	*15*
Vegetable Sales	*4*	*20*
Livestock Sales	*3*	*15*
Social Welfare Benefits	*5*	*25*
Other (bush removal)	*1*	*5*

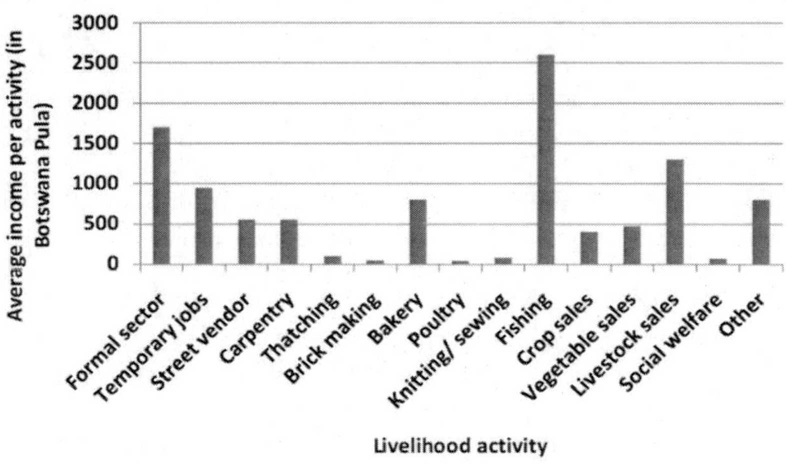

Figure 16. Average monthly income per livelihood activity (Botswana Pula).

Table 14. Summary of monthly expenditure of fishers from the Boteti River system

Frequency of Monthly Expenditures by Category (N=19)		*Proportion (%)*
Food	*19*	*95*
Alcohol/Tobacco	*7*	*35*
Clothing/Footwear	*11*	*55*
Transport	*16*	*80*
Education	*13*	*65*
Upkeep of fishing gear/employment	*9*	*45*
Other	*9*	*45*

However, frequency of monthly expenditures does not give a more complete picture than amount of money spent in each of the categories. While the majority of fishers spend their money on food (Table 15), the actual amount spent is just about half of what is re-invested into fishing (Figure 17). Clothing and footwear also take up a large proportion of total monthly expenditure, while alcohol and tobacco are relatively insignificant.

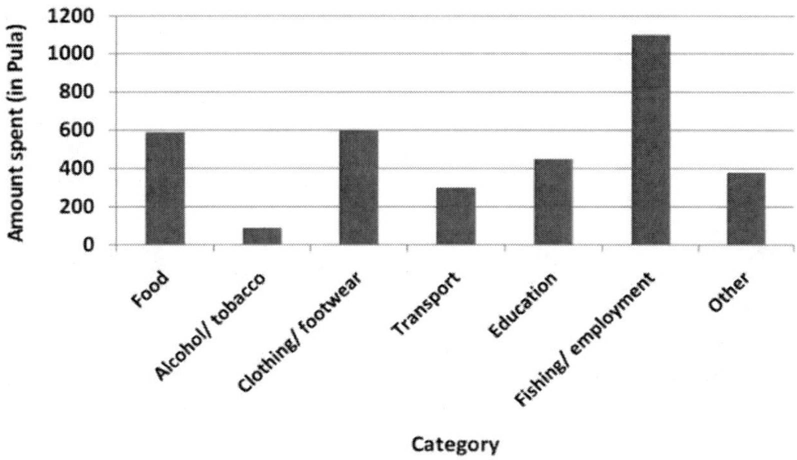

Figure 17. Average monthly expenditure by category (Botswana Pula).

Table 15. Summary of the importance of fish as a component of security in the fisher households from Boteti River

Food security	Proportion (%)
Fish as food source	
Fish makes up more than half of diet	47
Fish makes at least half the diet	53
Purchasing power	
Ability to purchase more than half of the food consumed	79
Ability to produce more than half of the food consumed	21
Ability to barter fish	
Ability to barter when necessary	63
Ability to barter occasionally	32
Inability to barter at all	5

Fish As a Source of Food Security

Fish is an important source of food security because it contributes at least half the diet to the majority (53%) of fishing households in Boteti River. That notwithstanding, a substantial proportion of fisher households also indicated that fish contributes to more than half their diet. A substantial proportion (79%) of the commercial households indicated that fishing gives them the ability to purchase more than half the food they consume. Moreover, the majority of commercial fishers (63%) indicated that they can also barter fish for other products when necessary (Table 14).

Furthermore, the first major strategy that commercial fishers undertake during times of food scarcity was to increase fish catches (69%), while a smaller proportion indicated that they would look for paid work as their major strategy (Figure 18). 32% of commercial fishers highlighted that increasing fish catches is their second major strategy during times of food scarcity, while 48% indicated that increasing fish catches is their third major strategy (Figure 18). The majority (78%) of fishing households perceive their children to be healthier than those from non-fishing households (Figure 19). This perception is based on the argument that their children have constant access to fish, which they perceive to be highly nutritious.

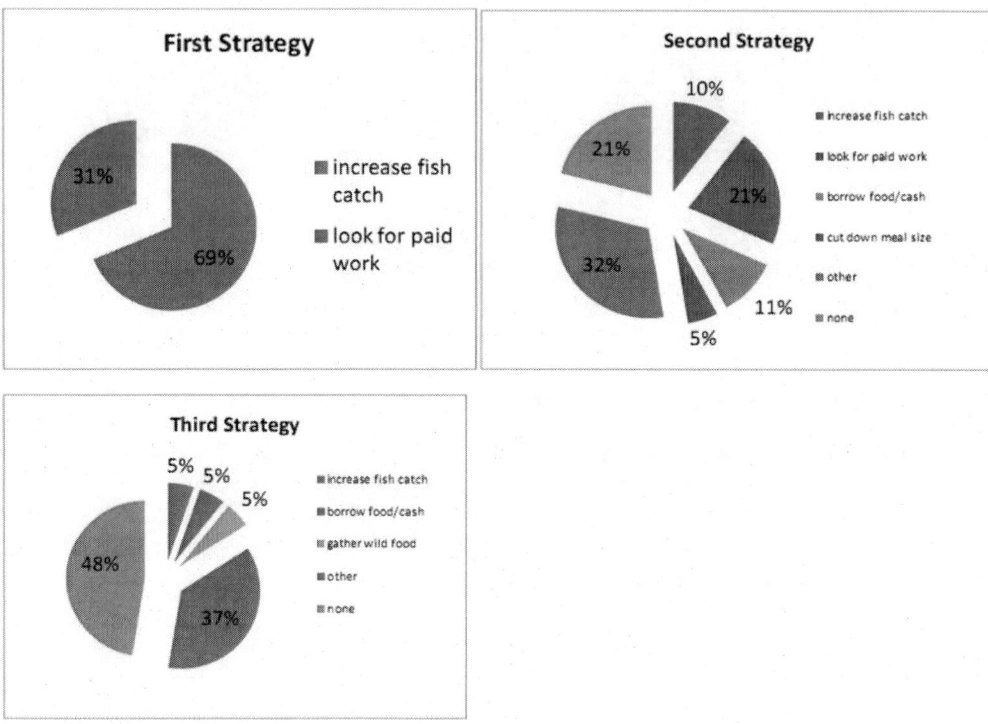

Figure 18. First, second and third strategies for fishing households during times of food scarcity.

Figure 19. Perceptions by fishing households on the health/ nutritional condition of children in their households compared to children from non-fishing households.

DISCUSSION

Fish Ecology and Biology

Past research from the Okavango Delta has produced evidence that the seasonal flood pulse is the main driver of ecological change in the Delta's fish community (Merron and Bruton, 1989; Merron, 1991; Mosepele et al., 2009, 2012). Merron and Bruton (1989) discuss that the annual flood re-connects water bodies that were previously disconnected and it also stimulates fish migrations in the Delta. It follows therefore, that the fish ecology in the Boteti River is also largely driven by the seasonal flood regime. Freshwater fish assemblages are diverse (Welcomme, 2011), which is similar to what was found in the Boteti River. The 39 fish species sampled in this river system compare very well with Merron's (1991) long-time series study in which 46 fish species were collected from Chanoga lagoon between 1983 and 1986. This observation suggests that speciation in this river system is very elastic taking into account that the Boteti River dried-up between the last time Merron (1991) conducted his surveys and this current study (a 25 year difference). Furthermore, the species check-list made here compares favourably with that of Skelton et al. (1985). However, Skelton et al. (1985) listed 56 species which is 5 more species more than sampled in this study. The main discrepancy for these two studies could be that the Skelton et al., (1985) check-list included Thamalakane River, while this study made a check-list for the Boteti River only.

While the most important species in the Boteti River according to this chapter was *B. lateralis*, this differs from Maar (1965) who observed that *M. macrolepidotus* was the most dominant fish species in the Boteti River system. Nonetheless, it is particularly noteworthy that the latter is still the third most important species in this river system. This suggests that while there might have been differences in sampling design between the current study and Maar (1965), and while there might indeed have been dramatic changes in the fish community driven (possibly) by changes in flooding patterns in the past 40+ years, there are still comparisons between the respective fish communities during these periods. Undoubtedly, this is an important ecological observation that highlights the resilience of fish populations in highly variable systems like the Boteti River. It also highlights the relative importance of fish refuges (e.g., Chanoga lagoon) which are then used to re-populate areas that were previously dry.

The significant correlation between sampling stations that are spatially closer (i.e., Xhobe vs. Chanoga and Chanoga vs. Motopi) also illustrates the dynamic fish migrations that are continuously occurring in systems such as these. Furthermore, this illustrates the existence of longitudinal migrations driven by seasonal flooding, as discussed by Merron and Bruton (1988) in the Delta. Furthermore, it is quite possible that the dynamic changes observed in diversity between the two sites were driven by *S. intermedius* and *M. macrolepidotus* which showed significant differences in catch rates between Xhobe and Chanoga. These dynamic changes are (possibly) driven by the flood regime, based on Mosepele et al.'s (2009) observations for the Okavango Delta; it is possible that these variations are driven by the flood regime. Maar (1965) also observed inter-annual variability in fish catch rates in Thamalakane River which were driven by different flood types in two years. Indeed these observations agree with Lowe-McConnell (1987) that fish movements in riverine fish communities (of floodplain systems) cause changes in their composition as a function of time and water levels.

Some dynamic observations were also done at the three experimental fishing stations along the Boteti River in this study. A progressive decrease in species diversity (based on the Shannon's diversity index), richness and evenness was observed along the river channel from Xhobe (which is upstream) to Chanoga until Motopi (which is further downstream). This spatial variation in species diversity could be caused by habitat partitioning in this system where upstream areas might have more diverse habitats than downstream habitats due to flooding duration. Here I argue that since upstream are flooded earlier, they might have had more time to support the development of micro-habitats than downstream habitats which flooded later. In that case, downstream habitats would be inhabited by pioneer species while upstream habitats would be characterized by a relatively climax fish community. This observation implies that fish migrations are driven by flooding patterns, and was first observed by Merron and Bruton (1984) in this river system.

It is particularly noteworthy that whereas *H. vittatus* was not sampled during this study, fishermen catch records revealed that 11 specimens were caught between 2007 and 2008. This is consistent with earlier studies (Maar, 1965; Mosepele and Mosepele, 2005) that while *H. vittatus* does not occur in this system, it periodically occurs during years of high floods.

The Role of Chanoga lagoon in the Boteti River

Chanoga is possibly the most important part of this river system due to an extensive and relatively deep lagoon that retains water longer than other parts of the Boteti River system (Merron and Bruton, 1984, 1988). Therefore, while this lagoon maintains a major fishery in the river system, it also plays a functional role as a fish refuge. Possibly, remnant populations of species in this river system reside in this lagoon during drought years and then assist in re-populating the river during years of high floods (and during extreme flooding events). This study has revealed that *B. lateralis* and *B. poechii* are the two most important species in this lagoon, while three cichlids are among the top ten most important species. This observation agrees with Merron et al., (1984) who also observed that *B. lateralis* dominated the fish community of this lagoon in a 1984 fish survey. Furthermore, they also found that *B. barnardi* and *B. fasciolatus* were also dominant species in this river system. These observed differences could have been caused by differences in sampling frequency and gear. Merron et al., (1984) used a seine net and rotenone while standardised experimental fishing nets were used in this study. Nonetheless, these observations illustrate that fish species succession in this system is very dynamic, and community structure can persist at different temporal and spatial scales.

Feeding Ecology and Speciation

According to Lowe-McConnel (1987), fish species in pulsing systems have plastic food items which is controlled by season, abundance of food item, "activity of fish" and change of habitat. Furthermore, Lowe-McConnel (1987) highlights that ubiquitous species such as *S. intermedius* have a generalist feeding behaviour compared to specialists whose distribution is relatively more restricted. While mosquito larvae was among the least important prey items in the Boteti River, its presence still agrees with Maar (1965) who observed that it is an important food source for the Delta's fish species. While it might have a diverse diet, *S. intermedius* is a strong piscivore and this has been observed before in the Okavango Delta by

Mosepele et al., (2005). The diverse food sources of this species in the Boteti River indicates that *S. intermedius* is a generalist predator that takes advantage of available food sources from both terrestrial and aquatic sources (Mosepele et al., 2005).

Interestingly, Squeakers (Synodontis spp) occur in very low numbers in this system, compared to upper panhandle habitats as observed by Merron and Bruton (1988). The most abundant cichlid (these are important food species in most African freshwater fisheries) in this system was the Banded tilapia (*T. sparrmanii*) while the least abundant cichlids were Nembwe (*S. robustus*) and Pink bream (*S. giardi*). These species occurrences make sense because the Banded tilapia is more of a habitat generalist and prefers relatively shallow and slow moving water while the other two species (Nembwe and Pink bream) prefer deep water habitats (Skelton, 2001). Furthermore, Banded tilapia has an omnivorous feeding ecology (generalist/ opportunistic feeder) while Nembwe prefers squeakers (which occur in low abundance already) and Pink happy prefers snails and bivalves which also possibly occur at low abundances. Interestingly, a top fish predator in the absence of tiger-fish (*H. vittatus*), the African pike (*H. odoe*), also occurs at very low abundance in this system. The most abundant predators appear to be the silver catfish (*S. intermedius*) and Purple-face large-mouth tilapia (*S. macrocephalus*).

The striped robber (*B. lateralis*) was the most important fish species in the Boteti River, while the dash-tail barb (Barbus poechii) was the second most important species. This is interesting because these two species are well known con-specifics in the Delta's panhandle (Merron and Bruton, 1988; Skelton, 2001) but this close relationship has never been recorded in a terminal river system before. Apart from the banded tilapia (*T. sparrmanii*) which is an omnivore, all the top five most species in this river system are insectivores. This observation suggests that insects (both terrestrial and aquatic sources) are a major source of energy into the aquatic system. Therefore, it is possible that the terrestrial system is actively subsidizing the aquatic environment, driven by seasonal flooding, which has been observed before in the Okavango Delta (Mosepele et al., 2012) other flood pulsing systems (Jardine et al., 2012). In these dynamic environments, most fish species are bound to have a generalist feeding behaviour, where they take advantage of any available food items. All the key species in this study show dynamic temporal changes in their feeding behaviour which attests to their generalist feeding behaviour. Notwithstanding, this perceived lack of diet specialisation suggests that the fish species present in this system are highly resilient to perturbations and may persist longer under variable environmental conditions. Mosepele et al., (2009) highlighted that the length of time water present and the nature of its flow determines fish community structure in the Delta. Whereas food might have some effect on habitat preferences, flooding is ultimately the main driver of change in the fish community.

Fishery Characteristics

There are temporal variations in fish catch rates from the gill net fishery which has been observed for the Okavango Delta by Mosepele (2000). This observation agrees with Maar (1965) who also observed temporal variations in experimental catch rates for the Thamalakane River. Nonetheless, the importance of tilapia (or bream) in the commercial fisher catches is testament to market preferences which have been observed in the Okavango Delta before (Mosepele, 2000; Mmopelwa et al., 2005; Kgathi et al., 2005; Mosepele et al., 2007; Mendelsohn et al., 2010). While Mosepele et al., (2007) observed that commercial fishers apply indigenous traditional knowledge to target their preferred species; it is equally possible that mesh selectivity in the fishery would also exploit a substantial portion of their preferred species as was observed in the Bangweulu swamps fishery (Kolding et al., 1996). Mean catch per commercial fisher in Boteti River is estimated at 1.14 tons yr^{-1} which is substantially lower than the 7.5 tons yr^{-1} that Cerdeira et al., (2000) estimated for Amazon commercial fishers. This might suggest that Boteti River fishers are much less efficient than their counterparts in the Amazon.

The species composition of the gill net catches also indicate that they are selective to only larger sized fish species, because of the low catches of *S. intermedius*. This is based on Merron and Bruton (1989) observation that *S. intermedius* is generally harvested using gill nets of mesh 50 – 60 mm. Conversely, fishers (in Boteti River) generally use gill net of mesh size range of 100 – 125 mm stretched mesh as described by Mosepele (2008). Therefore, while it represents an important biomass in the river system, it still remains relatively under-exploited. Equally important to note is that a much smaller species, *B. lateralis*, is the most dominant in this river system, and yet it is relatively under exploited. Furthermore, what is particularly interesting are high catfish catches (in tonnes) recorded from commercial gill net catches in this system. The absence of traditional fishing (i.e., fishing baskets, mosquito nets, fishing traps, etc.) in this river system, which have been observed in the Okavango Delta (Mmopelwa et al. 2009), suggests that there is a large proportion of the fish community that is unexploited. According to Mosepele et al., (2003), the key species exploited by fishing baskets are *T. sparrmanii* (a smaller sized cichlid), while topminnows (e.g., *A. johnstoni*) are mainly exploited by mosquito nets. Due to the dynamic nature of this ecosystem, it follows that there are always new fishers entering the fishery, who might not necessarily have an indigenous knowledge e of fishing. It is perhaps due to this that there is currently no traditional fishing in the Boteti River.

Socio-Economic Aspects of the Fishery

While the majority of fishermen are Ba-Yei (35%), it is interesting to note that other groups (BaKalanga, Namibian and Angolans) that have never been observed before in the Delta (i.e., upstream) are also present in the fishery. The Namibian and Angolan fishers indicated that they came from the Dukwe refugee camp, which is approximately 300 km away from this river system. Commercial fishing households resemble other households in the region with an average of 5 people per household (Mosepele, 2001). The majority (55%) of households have young children (i.e., 0-5 year olds), while only a small proportion of commercial fisher households (25%) have people older people (i.e., people aged 55 and older). Not surprisingly, the majority of commercial fishers had been fishing for only 1-2 years (the period when the river has been flowing), except for one old fisher who has been fishing since the 1970s. This suggests that commercial fishing in Boteti River is a relatively young activity, with (possibly) limited indigenous knowledge on fish harvesting techniques compared to fishing in the Delta, where there is a tremendous amount of indigenous traditional knowledge among relatively old commercial fishers as observed by Mosepele et al., (2007).

The socio-economic assets of commercial fishers in Boteti River suggests that they also have a strong inclination to arable and pastoral farming, similar to what has been observed in the Okavango Delta (Mmopelwa et al. 2005; Mosepele and Ngwenya, 2010). The strong interplay of agricultural activities with fishing enterprises attests to the importance of fishing as a social safety-net for households in the Boteti River. Furthermore, the commercial fishers in the Boteti River may be more adaptable to external shocks in the fishery because they have already experienced it when the river was dry for some time.

Fishing is a major source of household income where about 75% of commercial fishers indicated that fishing contributes well over half of monthly household income. In addition, the majority of average monthly expenditure (≈US$165.66) is also re-invested into fishing through rental fees (of fishing equipment), employment, and other monthly costs. That notwithstanding, while an average fisher generates income of approximately over US$376.51/ month, this compares very favourably with an average annual income of US$389.94/ fisher (Note: US$1 = BWP1.59 using 1987 rates) that was estimated in 1986/87 by Southern African Development Community [SADC] (1989). Furthermore, the SADC (1989) report highlights that the amount of income generated by fishing during this time was too low for fishers to re-invest it into their fishing activities, a situation completely different from the present. The relatively high incomes generated by commercial fishers in this system can be attributed to the absence of out-board engines, which would have contributed to high running costs. This observation was made by Camargo and Petrere (2001) in their analysis of the Sao Francisco River commercial fishery.

Fish is an important source of food security for commercial fishers' households in this river system. It is also an important source of high value protein similar to what has been observed in the Amazon by Cerdeira et al., (2000). The biggest coping strategy that commercial fisher households adopted during times of food scarcity was to increase fish catches as a first, second and third strategy. A similar observation was made in the Okavango Delta where Mosepele et al., (2006) highlighted that increasing fish catches was the commonest coping strategy in commercial fisher households. This buttresses the argument that fishing is a social safety net that fishers turn to when general macro-economic conditions are bad (Mosepele, 2000). It follows therefore, that any negative impact of fish production in this river would invariably have a negative impact on the livelihoods of the riparian community.

Extreme Flooding Events and Fisheries Dynamics

Generally, short term extreme flooding events can have negative impacts on fish populations as shown in the results. These include disturbing fish larval development which may result in high mortalities (Capra et al., 2003) especially for indicator 1 species such as the large tilapia (e.g., Oreochromis spp, *T. rendalli*) which are key commercial species (Mosepele et al., 2007; Mmopelwa et al., 2009) and *C. gariepinus* which is more valuable to the subsistence fishery (Mosepele et al., 2009). Furthermore, Arthington et al., (2003) highlights that extreme flooding can increase sedimentation which would either smother fish eggs or invertebrate prey. Given this particular scenario, both indicator 1 and 2 species would be negatively affected by extreme flooding due to loss of prey items. This is because aquatic invertebrates (especially chironomids) are key fish food items in the Boteti River (see Table 5) and loss of food would have a severely negative effect on fish growth and production. Furthermore, Welcomme and Halls (2004) observe that "rapid increases in water level" can submerge nests of bottom breeding species (e.g., Oreochromis spp and Tilapia spp) to a much greater depth which may result in spawning and hence recruitment failure. Moreover, Welcomme and Halls (2004) highlight that rapid increases in water levels can wash away eggs of marginal habitat breeding/spawning species (indicator 2) which would also result in breeding and hence recruitment failure.

Results have also revealed that a shallow hydrograph at transition 2 (which would normally occur after an extreme flood) for both two indicators and higher than normal floods for indicator 1 are generally beneficial to fish stocks. Welcomme and Halls (2004) highlight that a longer duration flooding regime makes available more food to most fish species and allows them to grow larger and generally enhance body condition. This intimate relationship between floodplain fish feeding ecology and the flood pulse has been established in the Okavango Delta (Lindholm et al., 2007; Mosepele et al., 2012) and other flood pulsed systems (Merona and Merona, 2004; Adite et al., 2005). Moreover, de Graaf (2003) showed that an extended flooding season (which is sometimes caused by extreme floods) sometimes results in two year classes which invariably enhances fish biomass production.

Despite the risks associated with extreme flooding in the Boteti River system, local fishermen have developed adaptation strategies to deal with environmental variability. Mmopelwa et al., (2009) observes that local fishers use different kinds of fishing gear to exploit fish species at different times of the year and in different fishing habitats. Therefore, if extreme floods increase the mean depth of their fishing habitats, fishers would generally use either gill nets or hook and line or both gears. Baran et al., (2009) discuss that diverse livelihood activities allow riparian communities to adapt to changes in fish availability. Therefore, the diverse livelihood activities identified in the Boteti River fishery (table 11) allow fishers to adapt to variability in fish production and availability. If extreme floods cause recruitment failure and hence low productivity, then fishers can/ will resort to other economic options, just like they did when the river had dried up for several years.

CONCLUSION

Flooding can have both negative and positive impacts on fish production in floodplain systems. This chapter has shown that fish production in the system is driven by seasonal flooding, and the fish community is highly resilient and well adapted to its high variability. Similar observations have been made for floodplain fisheries elsewhere (Craig et al., 2004). Poor or low flooding causes desiccation in seasonally flooded systems which results in high fish mortalities (Welcomme, 1986; Chapman and Kramer, 1991). Similarly, severe floods can wash away fish larvae and juveniles hence resulting in loss of a year class strength (Chapman and Kramer, 1991). This is a possible scenario that can occur in the Delta (and Boteti River) due to varying levels of upstream developments (Mosepele, 2009). High fish mortalities can occur in low water refuges when oxygen and food availability become limiting (Chapman and Kramer, 1991). According to Welcomme (1986), there is a threshold at which sustained high floods may not result in any further increases in fish biomass production while periods of low floods can have catastrophic effects on fish production. Therefore, at a local scale, while severe floods can have a negative impact on some species like poecilliids (Chapman and Kramer, 1991), they can also open up new channels into some permanent water bodies (lagoons) and hence aid in fish dispersal (Mosepele et al., 2009).

Chapman and Kramer (1991) observe that floods drive changes in fish density which subsequently changes the competitive environment, and changes the population structure and composition. Chapman and Chapman (1993) observed that the seasonal flood drives fish migrations between newly opened habitats. This seasonal flooding opens up new habitats and re-connects previously disconnected habitats at low flood levels (Mosepele et al., 2009). They (Chapman and Chapman, 1993) also observed that the seasonal flood results in annual fish migrations which cause dynamic changes in the fish community structure. Essentially, the slope of the hydrograph determines the probability of persistence of floodplain fish species abundance. In this case, Chapman and Chapman (1993) observed that a sharp slope of the hydrograph at the end of the flooding season results in high fish mortality because fish get trapped in the drying up floodplains.

High and sustained floods in floodplain systems can also have a positive impact on fish production (see Table 11), which is based on Junk et al.'s (1989) flood pulse concept, that ties floodplain fish production to the extent of the flooded area. In this situation, fish populations migrate with the floods due to improved food availability at the flood front (DeAngelis et al., 2010). Therefore flooding in seasonal floodplains is a key process that connects the terrestrial and aquatic environment (Mosepele et al., 2012), to produce food (i.e., fish) for higher trophic levels (DeAngelis et al., 2010). Studies from elsewhere (de Graaf, 2003) have shown that an extended flood in floodplain systems can result in more than one cohort which improves the year class strength of the fish population. This contributes to high fish production during years of good/high floods (Welcomme, 1986) which is essential for provision of nutrition and food security for riparian communities. Welcomme (1986, 1999) observes that the time at which this new biomass enters the fishery (observed as the response of the exploited fishery to flooding) under this scenario is an indicator of the status of the fishery, where new fish biomass is exploited much later in lightly fished systems and the reverse is true for intensely fished systems. It is within this background that Jardine et al., (2012) advocates for a more comprehensive understanding of the terrestrial – aquatic linkages in floodplain fisheries, so that management of these systems is much more comprehensive that it currently is.

This chapter has shown that the gill net fishery (based on catch and effort data) is a major source of livelihoods for the riparian (fishing) community along the Boteti River system. Mosepele et al., (2006) made a similar observation that the Okavango Delta's fishery is a major source of food security for its riparian communities. Ngwenya and Mosepele (2007) also observed that fish is a major source of nutrition for the chronically ill in the Okavango Delta, while Nnyepi et al., (2007) highlighted that it is a major source of nutrition for children from fishing families. Indeed river fisheries underpin the livelihoods of many riparian communities in systems such as The Amazon, Niger, The Mekong, Congo, and the Ganges (Neiland and Bene, 2003; Dugan et al., 2010) among others. Other studies have shown that small sized fish (species) are a major source of key nutrients (e.g., Iron, Calcium, Amino acids, Calcium, etc) that are valuable to women and children under the age of 5 (Larsen et al. 2000; Roos et al., 2003; Baran, et al., 2007; Dugan et al., 2010). Encouraging the development of a fishery targeted at these smaller sized species might unlock their potential to the river's riparian community. Therefore, this chapter has shown that while floods might generally have negative impacts on society, they also have beneficial effects that can sustain livelihoods of riparian communities.

REFERENCES

Allan, D., Abell, R., Hogan, Z., Revenga, C., Taylor, B.W., Welcomme, R.L., Winemiller, K., (2005). Over-fishing of Inland Waters. *BioScience,* 55(12): 1041-1051.

Arthington, A.H., Rall, J.L., Kennard, M.J., Pusey, B.J., (2003). environmental flow requirements of fish in Lesotho rivers using the drift methodology. *River Research and Applications*, 19: 641–666.

Baran E., Jantunen T., Chong C. K., (2007). Values of inland fisheries in the Mekong River Basin. World Fish Center, Phnom Penh, Cambodia.

Barnes, J., Saraiva, R., Mmopelwa, G., Mbaiwa, J., Magole, L., Wamunyima, D., (2009). Okavango river basin trans-boundary diagnostic analysis: Socio-economic assessment. OKACOM, Maun, Botswana.

Berkes, F., (2003). Alternatives to conventional management. *Environments,* 31(1), 5-19.

Camargo, S.A., Petrere, M., (2001). Social and financial aspects of the artisanal fisheries of Middle Sao Francisco River, Minas Gerais, Brazil. *Fisheries Management and Ecology*, 8, 163 -171

Capra, H., Sabaton, C., Gouraud, V., Souchon, Y., Lim, P., (2003). A population dynamics model and habitat simulation as a tool to predict brown trout demography in natural and bypassed stream reaches. *River Research and Applications*, 19, 551–568.

Cerdeira, R.G.P, Ruffino, M.L., Isaac, V.J., (2000). Fish catches among riverside communities around Lago Grande de Monte Alegre, Lower Amazon, Brazil. *Fisheries Management and Ecology*, 7, 355 - 374.

Chapman, L.J., Kramer, D.L., (1991). The consequences of flooding for the dispersal and fate of poecillid fish in an intermittent tropical stream. *Oecologia*, 87, 299-306.

Chapman, LJ. Chapman, C.A., (1993). Fish populations in tropical rain pools: A re-evaluation of Holden's data on the river Sokoto. *Ecology of Freshwater Fish*, 2, 23 – 30.

DeAngelis, D.L., Trexlerc, J.C., Cosnerd, C., Obazac, A., Jopp, F., (2010). Fish population dynamics in a seasonally varying wetland. *Ecological Modelling* 221, 1131–1137.

Delgado, C.L., Wada, N., Rosegrant, M.W., Meijer, S., Ahmed, M., (2003). Outlook for fish to 2020: meeting global demand. *World Fish Center*, Malaysia.

Dugan, P., Delaporte, A., Andrew, N., O'Keefe, M., Welcomme, R., (2010). Blue Harvest: Inland Fisheries as an Ecosystem Service. *World Fish Center*, Penang, Malaysia.

Food and Agricultural Organisation., (2005). Increasing the contribution of small-scale fisheries to poverty alleviation and food security. Rome, Italy.

Gumbricht, T., McCarthy, J., McCarthy, T.S., (2004). Channels, wetlands and islands in the Okavango Delta, Botswana, and their relation to hydrological and sedimentological processes. *Earth Surface Processes and Landforms*, 29(1): 15 – 29.

Hoggarth, D.D., Cowan, V.J., Halls, A.S., McGregor, M., Garaway, C.A., Payne, A.I., Welcomme, R.L., (1999). Management guidelines for Asian floodplain river fisheries. Part 1. A spatial, hierarchical and integrated strategy for adaptive co-management. *FAO Fisheries Technical Paper*. No. 384/1. Rome, Italy.

Jardine, T.D., Pusey, J.B., Hamilton, S.K., Pettit, N.E., Davies, P.M., Douglas, M.M., Sinnamon, V., Halliday, I.A. Bunn, S.E., (2012). Fish mediate high food web connectivity in the lower reaches of a tropical floodplain river. *Oecologia*, 168, 829–838.

Kgathi, D.L., Mmopelwa, G., Mosepele, K., (2005). Natural Resource Assessment in the Okavango Delta, Botswana; Case Studies of Some Key Resources. *Natural Resource Forum*, 29: 70 - 81.

King, J., Brown, C., Sabet, H., (2003). A scenario-based holistic approach to environmental flow assessments for rivers. *River Research and Applications*. 19, 619–639.

King, J.M., Brown, C.A., (2009). Scenario Report (ecological and social–climate change) Volume 3 of 4. Report 07-2009 EPSMO/BIOKAVANGO Okavango Basin Environmental Flows Assessment Project, OKACOM, Maun, Botswana.

Kolding, J and Paul van Zwieten, P.A.M., 2012. Relative lake level fluctuations and their influence on productivity and resilience in tropical lakes and reservoirs. *Fisheries Research*, 115– 116, 99– 109.

Kolding, J., Ticheler, H., Chanda, B., (1996). Assessment of the Bangweulu swamps fisheries. WWF Bangweulu Wetlands Project.

Larsen, T., Thilsted, S.H., Kongsbak, K., Hansen, M., (2000). Whole small fish as a rich calcium source. *British Journal of Nutrition*, 83: 191–196.

Maar, A., (1965). Report on fisheries survey in Bechuanaland B.P in the years 1963/64.

Mendelsohn, J.M., VanderPost C., Ramberg, L., Murray-Hudson, M., Wolski, P., Mosepele, K., (2010). Okavango Delta: floods of life. RAISON (Windhoek, Namibia).

Merron, G.S., (1991). The ecology and management of the fishes of the Okavango Delta, Botswana, with particular reference to the role of the seasonal floods. *Rhodes University, PhD thesis*. Pp. 171.

Merron, G.S. Bruton, M.N., (1984). Report on the June-July 1984 expedition to the Okavango Delta, Botswana. J L B Smith Institute of Ichthyology, *Investigational Report*, No. 12.

Merron, G.S. Bruton, M.N., (1989). Recent fisheries research in the Okavango Delta. *South African Journal of Science*, 85: 416 – 417.

Mmopelwa, G., Mosepele, K., Mosepele, B., Moleele, N., Ngwenya, B., (2009). Environmental variability and the fishery dynamics of the Okavango Delta, Botswana: The case of subsistence fishing. *African Journal of Ecology,* 47(1): 1–9.

Mmopelwa, G., Raletsatsi, S., Mosepele, K., (2005). Cost Benefit Analysis of Commercial Fishing in Shakawe, Ngamiland, Botswana. *Botswana Notes and Records*, 37: 11–21.

Mosepele K., Mosepele B., Wolski P., Kolding J., (2012). Dynamics of the feeding ecology of selected fish species from the Okavango River delta, Botswana. *Acta Ichthyologica et Piscatoria*, 42 (4): 271 – 289.

Mosepele, K., (2000). Preliminary length based stock assessment of the main exploited stocks of the Okavango delta fishery. MPhil Thesis, Department of Fisheries and Marine Biology. University of Bergen, Norway.

Mosepele, K and Mosepele, B., (2005). Spatial and Temporal Variability in Fishery and Fish Community Structure in the Okavango Delta, Botswana: Implications towards fisheries management. *Botswana Notes and Records*, 37: 280 – 291.

Mosepele, K., Ngwenya, B., (2010). Socio-economic survey of commercial fishing in the Okavango Delta, Botswana. Bay Publishing: Gaborone, Botswana.

Mosepele, K., Ngwenya, B., (2010). Socio-economic survey of commercial fishing in the Okavango Delta, Botswana. (Bay Publishing: Gaborone, Botswana).

Mosepele, K., (2008). Flood pulse in a sub-tropical floodplain fishery and the consequences for steady state management. In: Totolo, O (Editor). Proceedings of the 2nd IASTED

Conference on Water Resource Management; Science and technology innovation for sustainable development. Calgary, Canada, pp. 56-62.

Mosepele, K., (2009). Okavango River Basin Technical Diagnostic Analysis: Environmental Flow Module Specialist Report Country: Botswana. Discipline: Fish. OKACOM, Maun, Botswana.

Mosepele, K., Mmopelwa, G, Mosepele, B., Kgathi, D. L., (2007). Indigenous knowledge and fish utilization in the Okavango Delta, Botswana: Implications for food security. In: Ahmed, A (Ed.). Managing knowledge, technology and development in the era of information revolution. Griffith University, Australia pp. 292 - 302.

Mosepele, K., Mosepele, B., Wolski, P. Kolding, J., (2012). Dynamics of the feeding ecology of selected fish species from the Okavango delta, Botswana. *Acta Ichthyologica et Piscatoria*, 42 (4): 271 – 289.

Mosepele, K., Moyle, P.B., Merron, G.S., Purkey, D. Mosepele, B., (2009). Fish, floods and ecosystem engineers; interactions and conservation in the Okavango Delta, Botswana. *Bioscience*, 59 (1): 53–64.

Mosepele, K., Ngwenya, B.N. Bernard, T., (2006). Artisanal fishing and food security in the Okavango Delta, Botswana, pages 159 – 168. In: Ahmed, A (Editor). Global and local resources in achieving sustainable development. (Interdescience Publishers: Geneva, Switzerland).

Mosepele, K., Williams, L., Mosepele, B., (2005). Assessment of the feeding ecology of the Silver Catfish (Schilbe intermedius, Ruppel, 1832) in a seasonal floodplain of the Okavango Delta, Botswana. *Botswana Notes and Records,* 37, 208 – 217.

Neiland, A. Béné, C., (2003). Review of River Fisheries Valuation in West and Central Africa. The World Fish Center, Malaysia.

NEPAD, (2005). The NEPAD action plan for the development of African fisheries and aquaculture.

Ngwenya, B., Mosepele, K., (2007). HIV/ AIDS, artisanal fishing and food security in the Okavango Delta, Botswana. *Physics and Chemistry of the Earth,* 32: 1339–1349.

Nnyepi, M., Ngwenya, B. Mosepele, K., (2007). Food (in) security and child nutrition in Ngamiland, pp. 281 - 291. In Ahmed, A (Ed.). Managing knowledge, technology and development in the era of information revolution. Griffith University, Australia.

Petrere, M., Welcomme, R.L. Payne, A.I., (1998). Comparing river basins world-wide and contrasting inland fisheries in Africa and Central Amazonia. *Fisheries Management and Ecology*, 5: 97-106.

Poff, N.L. Allan, J.D., (1995). Functional organization of sream fish assemblages in relation to hydrological variability. *Ecology*, 76(2): 606-627.

Poff, N.L., (1992). Regional hydrologic response to climate change: an ecological perspective pp. 88 – 115. In: P.L. Firth and S.G. Fisher (Editors). Global climate change and freshwater ecosystems. Springer-Verlag, New York, USA.

Roos, N., Islamy, M.M., Thilsted, S.H., (2003). Small Indigenous Fish Species in Bangladesh: Contribution to Vitamin A, Calcium and Iron Intakes. *Journal of Nutrition, Supplement*, 4021S – 4026S.

Southern African Development Community [SADC]., (1989). Regional fisheries survey; SADC country situation report.

Skelton, P.H., Bruton, M.N., Merron, G.S., van der Waal, B.C.W., (1985). The fishes of the Okavango drainage system in Angola, South West Africa and Botswana: Taxonomy and distribution. *Ichthyological Bulletin*, No. 50.

Smith, L.C., Alderman, H. Aduayom, D., (2006). Food Insecurity in Sub-Saharan Africa; New Estimates from Household Expenditure Surveys. Research Report 146. *International Food Policy Research Institute*, Washington, D.C.

Tichler, H.J. Chanda, B., (1993). Frame survey of the Bangweulu fishery, 1992. Department of Fisheries, Report No. 4.

Welcomme R.L., Halls A., (2004). Dependence of Tropical River Fisheries on Flow, pp 267-284. In: Welcomme, R.L and Petr, T (Eds), Proceedings of the second international symposium on the management of large rivers for fisheries, volume 2: Sustaining Livelihoods and Biodiversity in the New Millennium. *FAO and the Mekong River Commission*.

Welcomme, R., (2011). Review of the state of the world fishery resources: inland fisheries. FAO, Rome.

Welcomme, R.L., (1986). The effects of the Sahelian drought on the fishery of the central delta of the Niger River. *Aquaculture and Fisheries Management,* 17: 147-154.

Welcomme, R.L., (1999). Review of a model for qualitative evaluation of exploitation levels in multi-species fisheries. *Journal of Fisheries Ecology and Management*, 6: 1-20.

In: Flooding
Editors: M. R. Motsholapheko and D. L. Kgathi

ISBN: 978-1-63117-607-4

Chapter 11

TOWARDS A STRATEGY FOR FLOOD RISK MANAGEMENT

D. L. Kgathi [*] ***and M. R. Motsholapheko***
Okavango Research Institute, University of Botswana

This book highlights the international experience of flood risk in five countries of Argentina, Australia, Botswana, Brazil, and Taiwan. It demonstrates that flooding is one of the worst hazards in both the developed and developing world. Floods are rated the third most common natural disaster after storms and earthquakes (World Bank/United Nations, 2010). Despite the global efforts to manage floods and mitigate their damages, there is evidence that the losses caused by floods continue to increase (Sayers *et al*., 2013). The book indicates that flooding is not only caused by too much water but results from a number of conditions. These include extreme meteorological events and poor land-use planning (Chapter 2). Based on the predictions of IPCC (2012), extreme weather events are expected to increase in the future as a result of climate change and this implies that they are likely to exacerbate the damages resulting from floods. Apart from these challenges, floods have also created opportunities as they enrich the land for agriculture by depositing nutrient-rich sediments in floodplains and therefore contribute to the provision of ecosystem services (Chapter 10; Kgathi *et al.,* 2013). Since the 1980s, many countries realised need to live in harmony with the processes of flooding as it is almost impossible to eliminate them. The book supports the view of Sayers et al. (2013) that there is no single solution for addressing floods and that it is necessary to develop a portfolio of measures. Risk-based approaches are recommended as a strategy for addressing the challenges of flooding.

In Australia, flood damage was estimated at $20 billion in 2010-12 and over 25 people were killed by floods. In the State of Queensland, 70% of the state was affected by floods and 75% of the banana crop was destroyed (Chapter 2). Flood risk assessments are necessary to undertake in order to manage floods and formulate mitigation plans and emergency strategies. In Australia, a rainfall-based method called the Design Event Approach is recommended for flood risk assessments. However, the main problem of this method is that the design rainfall

[*] Corresponding Author e-mail: dlkgathi@ori.ub.bw.

input variables are probability neutral since they are only based on stochastic nature of rainfall depth and does not include other stochastic flood producing input variables in the estimation of design flood. A new method known as the Joint Probability Approach, which takes into consideration the stochastic nature of flood variables, is being researched in Australia. The method has been applied to the Orara River catchment in the state of New South Wales and the results have produced accurate flood estimates. The method could be applied to other Australian States and other parts of the world with similar conditions.

A case study of flooding in north-east province of Buenos Aires in Argentina shows the two geomorphologic environments of the coastal and continental plains are vulnerable to flooding (Chapter 3). The main causes of flooding are modification of the drainage system by urbanisation and lack of suitable land-use planning. The lack of suitable land-use planning that takes into consideration the characteristics of the physical environment has led to the occupation of low-lying, flood prone areas, thus aggravating flooding. This has led to periodic floods with severe socio-economic and environmental impacts. The flash flood of 2nd April, 2013 in La Plata killed more than 100 people and also seriously damaged property. The chapter recommends that the impacts of flooding could be ameliorated by developing efficient monitoring system and introducing a series of mitigation measures.

Chapter 4 assessed the impacts of floods and evaluated the incidence of leptospirosis, a disease commonly caused by infection with *leptospira,* in various geographic locations of the São Paulo Metropolitan Area (SPMA) in Brazil. Through the use of GIS techniques a map of areas vulnerable to leptospirosis was developed. The chapter revealed an increasing incidence of the disease which was also linked to an increase in precipitation. Through the development of a vulnerability map, areas of high vulnerability could be identified and proper interventions undertaken. The development of vulnerability maps as part of flood risk assessments is important for guiding land-use policies and for developing communication strategies in flood-prone areas. Although the levels of precipitation and subsequent flooding cannot be controlled, knowledge of areas of high risk of infection with waterborne diseases can help in reducing contact with animal excretions and preventing contamination of potable water sources.

In Taiwan, flood risk is one of the major hazards, particularly in the Taipei metropolitan area, where about 80% of the population lives (Chapter 5). Most of the urban areas are located near large rivers as a result of the domestic and industrial demand for water. Structural measures still play an important role in decreasing flood risk in Taiwan. In order to mitigate protect the Taipei metropolitan area from the 200-year flood of the Tanshui River, a large scale flood prevention programme called the Taipei Flood Prevention System was introduced in 1963 and finalised in 1999. The programme included structural measures such as protection levees and dykes which were constructed along the river. The results of sensitivity analysis of thirty-two scenarios revealed that the Taipei Flood Prevention System is likely to face problems in the future as a result of urbanisation and climate change. For instance, these factors were predicted to increase the riverbed roughness by 25% and the 200-year return period flood discharge by 13%. As a result the Taipei Flood Prevention programme, based on levies and pumps to prevent flooding, will face challenges in the future. The chapter concluded that the efficiency challenges faced by this flood prevention system are useful lessons for other developing countries with similar programmes.

Access to insurance cover in flood-prone areas is one of the hotly debated measures for adapting to flood risk in the context of anthropogenic climate change (Wilby and Keenan,

2013). Chapter 6 discusses the extent to which the people of Queensland in Australia have access to flood risk insurance. It demonstrated that the 2011 floods adversely affected many households in the state of Queensland because they did not have flood insurance cover. Although the insurance policies for flood risk cover increased from 3% in 2006 to 54% in 2010 in Queensland, this proportion is much lower than in other states in Australia. The chapter makes an attempt to determine some of the subjective factors associated with the failure to insure for flood risk in Queensland. Using data from a survey of 501 residents of the three major cities of south-eastern Queensland (Australia), key subjective factors associated with non-insurance for flooding were analysed. These included risk perception and community attitude toward institutional arrangements for flood risk response. The chapter concluded that the majority of residents had no flood cover. This was partly attributed to moral hazard. For instance, those who expected to be compensated for disaster relief were less likely to insure for flooding, suggesting that their failure to insure could be due to the expectation of receiving disaster relief.

In Chapter 7, the book examines the pertinent issues of behavioural change amongst riparian communities amenable to flood risk in the Okavango Delta. The chapter explains why local people perceive and respond to flooding events in the way they did. The perceptions of and dispositions of community members on floods were discussed. The unveiling of their preferences and desires to stick to the known rather than the unknown in a bid to perpetuate certain livelihood strategies and cultural relations provides an opportunity to devise new paradigms and policy reforms in implementing people-centred flood risk intervention programmes. The chapter offers some insights on how the application of Kurt Lewin's 3-step model of planned change could be used to achieve household behavioural change for minimising flood risk.

In Botswana, extreme weather events have also been experienced in most parts of the country, including the Okavango Delta, where extreme flooding often occurs during the multi-decadal wet phase periods. Flooding adversely affected flood recession agriculture, livestock farming and also disrupted public infrastructure and displaced households (Chapter 8). In the recent wet phase which was noted to have started around 2004, extreme floods were experienced in the Okavango Delta. The financial costs of flood damage in Ngami sub-District of Ngamiland alone were estimated to be US$ 83 460 in 2010 and US$ 577 800 in 2011. In the Okavango sub-District, the flood damage costs were estimated at USA$ 231 560 in 2009 and USA$ 797 624 in 2012 (Motsholapheko et al., 2013). The book indicates that some households coped with extreme floods through labour switching, temporary local mobility, and enrolling in government assistance programmes. Other households (62%) autonomously adapted to extreme floods by diversifying their livelihood activities, resorting to local mobility, and migrating to other areas. The failure to adapt to flooding was associated with lack of access to various forms of capital. The chapter further indicates that there was a decline in access to human and social capital in the Okavango Delta in the last 30 years due to a number of factors such as economic transformation, migration of the able-bodied population to urban areas, disintegration of family structure, and collapse of indigenous institutions.

The book also demonstrates how communicating risk can reduce the effects of environmental hazards such as floods (Chapter 9). Specifically, the chapter explores the factors that contribute to low adoption of risk warning information on floods by the rural communities of the Okavango Delta. It indicates that low adoption of risk warning information is caused by a number of factors such as the time taken without experiencing floods, myths, beliefs and perceptions about floods. In some cases, the recipients of flood information wait for the floods to arrive in their area before they evacuate because they did not trust those who communicate the risk information. The chapter recommends that there is need for risk communicators to take into consideration the perceptions and views of local communities on flood risk before designing policy interventions for addressing this problem.

Through the use of environmental flow assessment methods, Chapter 10 determined the impacts of extreme flooding on fish in the Boteti River, an outflow channel of the Okavango Delta in Botswana. The chapter established that although short-term extreme flooding had some negative impacts including disturbance on fish larval development, prolonged flooding had positive impacts on fish productivity. Furthermore, the chapter established that this improved the well-being of fishing communities along the Boteti River. From these findings it can be adduced that floods do not necessarily have negative impacts, but also provide useful opportunities which can be beneficial for the affected communities. This finding further buttresses observations that flooding supports ecosystems and largely enhances the provision of ecosystem services that support livelihoods in wetlands around the world (Wisner *et al.*, 2004; Samuels *et al.*, 2006; Rebelo et al., 2010; Kgathi *et al.,* 2013). In many cases, particularly in wetlands, the benefits may be sufficient to offset the impacts of flooding. However, the ability to identify and use opportunities that arise from flood events has spatial, temporal, socio-economic and cultural differences. Such differences have been observed to result in institutional barriers to effective response to hazards including flooding (see Agrawal, 2008; Adger *et al.*, 2009)

The book has indicated that floods are associated with both challenges and benefits. Despite attempts to manage floods and mitigate their damages, the losses of floods continue to increase in both the developed and developing countries. On the basis of the lessons learnt from the case studies, the following policy directions need to be emphasised in order to prevent floods and mitigate their impacts:

- As a result of current challenges such as climate change, there is need to determine the effectiveness of the flood risk assessment methods from time to time. There is evidence that the traditional approaches for flood defence are not yielding sufficient results for managing flood risk (Wilby and Keenan, 2013).
- Vulnerability maps need to be developed as part of risk assessments; this can guide land-use policies and avoid exposure to waterborne diseases associated with flooding. Maps can also serve as important tools for risk communication as revealed by the case study of Brazil in this book.
- In order to reduce flood risk, both structural and non-structural measures should be adopted, and attempts should be made to monitor the impacts of these measures overtime. The case study of Tanshui River in Taiwan has revealed that structural measures were not effective in flood management and non-structural measures were recommended.
- There is a need to promote suitable land-use planning that considers the characteristics of the physical environment. This will guide policy makers in ensuring that people do not settle in flood-prone areas as this may aggravate the socio-economic impacts of flooding.

- Flood risk insurance should be promoted as one of the measures of flood risk reduction. This book has revealed that even in developed countries such as Australia, the majority of residents have no flood risk cover.
- Community level efforts to reduce the impacts of flooding should take cognisance of the prevailing socio-cultural conditions in order to circumvent societal barriers to risk avoidance.
- Participatory communication approaches should be adopted as one of the measures of flood risk communication as revealed by the study on the rural communities in the Okavango Delta, Botswana.
- Flood risk interventions should include institutional measures that aim to enhance the ability of affected communities to identify, use and adequately benefit from opportunities emanating from flooding and other hazards.

In conclusion, it is necessary to consider all the causes of flooding when formulating flood risk management policies. Although flooding adversely impacts on human well-being, some of its positive aspects such as improved biological productivity and enhanced ecosystem services can provide opportunities for improved human well-being, particularly among the poor and marginalised rural communities. It is therefore necessary to ensure that the prevention and mitigation of floods does not result in degradation of ecosystem services.

REFERENCES

Adger, W.N., Dessai, S., Goulden, M., Hulme, M., Lorenzoni, I., Nelson, D.R., Naess, L.O., Wolf, J., Wreford, A., (2009). Are there social limits to adaptation to climate change? *Climatic Change*, 93: 335-354.

Agrawal, A., (2008). The role of institutions in adaptation to climate change. International Forestry Research and Institutions Program (IFRI), *Working Paper*. Washington DC: World Bank.

Kgathi, D.L., Mmopelwa, G., Vanderpost, C., Wolski, P., Motsholapheko, M.R., (2013). Impacts of desiccation on ecosystem services and household adaptation in the Okavango Delta, Botswana. *Biodiversity and Ecology,* 5, 263-277. DOI: 210.7809/b-e.00280.

Motsholapheko, M.R., (2013). Rural livelihoods and household adaptation to flood variability in the Okavango Delta, Botswana. *PhD Thesis,* University of Botswana, Gaborone.

Motsholapheko, M.R., Kgathi, D.L., Vanderpost C., (2013). An assessment of adaptation planning for flood variability in the Okavango Delta, Botswana. *Mitigation and Adaptation Strategies for Global Change*, 18(6). DOI: 10.1007/s11027-013-9488-5.

Rebelo, L.M., McCartney, M., Finlayson, C., (2010). Wetlands of Sub-Saharan Africa: distribution and contribution of agriculture to livelihoods. *Wetlands Ecology and Management*, 18(5): 557-572.

Samuels, P., Klijn, F., Dukman, J., (2006). An analysis of the current practice of policies on river flood risk management in different countries. *Irrigation and Drainage*, 55: 5141-5150.

Sayers, P. Yuanyuan, L., Galloway, G., Fuxin, S., Kang, W., Yiwei, C., Quense, T.L., (2013). *Flood Risk Management: a strategic approach.* Paris, UNESCO.

Schanze, J., Zeman, E., Marsalek, J. (Eds), (2006). *Flood risk management- hazards, vulnerability and mitigation measures.* Berlin, Germany: Springer.

United Nations International Strategy for Disaster Reduction [UNISDR], (2011). *Global assessment report on disaster risk reduction: revealing risk, redefining development.* Geneva: United Nations.

Wilby R.L., Keenan, R., (2012). Adapting to flood risk under climate change. *Progress in Physical Geography*, 36(3): 348-378.

Wisner, B., Blaikie, P., Cannon, T., Davis, I., (2004). *At risk: natural hazards, people's vulnerability and disasters* (2nd Edition). London: Routledge.

World Bank/United Nations, (2010). *Natural unnatural disasters, the economics of effective prevention*. Washington DC: World Bank.

LIST OF CONTRIBUTORS

Wilfredo Caballero, PhD

Dr Wilfredo Caballero has completed his PhD in flood hydrology from University of Western Sydney in 2013, Australia. He has over 15 years' experience as a civil engineer. His research interests include application of joint probability approach to flood estimation, statistical hydrology and rainfall runoff modelling. He has produced ten scholarly publications in water and environmental engineering field.

Md Mahmudul Haque, MSc

Mr Md Mahmudul Haque obtained his B.Sc and MSc in civil engineering from Bangladesh University of Engineering and Technology (BUET). He is currently a PhD scholar at the University of Western Sydney, Australia. He has over 8 years' experience as a civil engineer. His research interests include flood frequency analysis, long term water demand forecasting, catchment yield and climate change impact analysis. He has published 14 research papers and book chapters in water and environmental engineering field.

Gwo-Wen Hwang, PhD

Dr. Gwo-Wen Hwang is an Assistant Research Fellow at the Hydrotech Research Institute at National Taiwan University (NTU); Adjunct Assistant Professor, St. Mary's Junior College of Medicine, Nursing and Management; Adjunct Assistant Research Fellow, Center for Weather Climate and Disaster Research at NTU. He earned his PhD from the Department of Civil Engineering, NTU, Taiwan. His research interests include river mechanics, sediment transport, ecological engineering, wetlands and river restoration. He has authored or co-authored over 9 peer-reviewed journal articles.

Donald L Kgathi, PhD

Dr Donald L Kgathi is a Professor of research in sustainable development at the Okavango Research Institute, University of Botswana. He holds a PhD in Development Studies (University of East Anglia, Norwich), MA in Development Economics (UN Institute for Economic Development and Planning, Dakar) and BA in Economics (University of Botswana). His current research interests are in economics of rural livelihoods, adaptation to the impacts of climate change, economics of ecosystem services, socio-economic impacts of biofuel production and use, and economic valuation of natural resources.

Oluwatoyin Dare Kolawole, PhD

Dr Oluwatoyin Dare Kolawole obtained a PhD (Rural Sociology & Agricultural Extension) and MA (Development Studies) from the Obafemi Awolowo University, Ile-Ife, Nigeria and Institute of Development Studies (IDS) at the University of Sussex in Brighton, United Kingdom, respectively. His main interest is in rural development and has worked extensively in the interface of science, policy and agriculture in Africa. He seeks to find a meeting point between science and local knowledge in the development process. Amongst others, he is a member of the Editorial Board of SAGE Open. He has been awarded the Visiting Canterbury Fellowship of the University of Canterbury in Christchurch, New Zealand for 2014.

Jihn-Sung Lai, PhD

Dr Jihn-Sung Lai is a Research Fellow at the Hydrotech Research Institute at National Taiwan University (NTU); Adjunct Professor, Department of Bioenvironmental Systems Engineering at NTU; Division Chief, Disaster-Prevention Project Division, Center for Weather Climate and Disaster Research at NTU. He earned his PhD from the Department of Civil and Environmental Engineering, University of California, Berkeley, USA. His research interests include river mechanics, flood inundation simulation, reservoir sedimentation and de-siltation, sediment transport, hydraulic physical and numerical modeling. He has authored or co-authored over 50 peer-reviewed journal articles.

Alex Lo, PhD

Dr Alex Lo is a political economist and geographer. He holds degrees in finance, environmental management, human geography, and political economy. He received his PhD from the Australian National University and is currently affiliated with the Urban Research Program at the Griffith University, Australia. His research focuses on people's risk perception and environmental attitude, community's economic adaptation to climate change, and the political economy of climate change.

Gagoitseope Mmopelwa, PhD

Dr Gagoitseope Mmopelwa is Senior Lecturer in the Department of Environmental Science, University of Botswana. He holds a PhD in Agricultural Economics from the University of Pretoria, MSc in Natural Resources Management from the Agricultural University of Norway and BSc in Agriculture from the University of Botswana. His research interests are in economic valuation of natural resources, economics of conservation, livelihood systems and adaptation to climate change.

Ketlhatlogile Mosepele, MSc

Mr Ketlhatlogile Mosepele is a Senior Research Scholar in Fisheries Biology and Management at the Okavango Research Institute of the University of Botswana. He holds MPhil in Fisheries Biology and Management from the University of Bergen, Norway, where he is also finalizing his PhD in the same field. His research interests are in the management of flood pulsed systems, the effect of the seasonal flood pulse on floodplain fish dynamics and the value of fisheries to riparian communities. He is currently working on flooding dynamics and water level fluctuations and their impact on freshwater fish biology in the Okavango Delta, Botswana.

Moseki Ronald Motsholapheko, PhD

Dr Moseki Ronald Motsholapheko is a Research Scholar in Water Resources Management at the Okavango Research Institute, University of Botswana. He holds a PhD in Natural Resources Management from the University of Botswana and a Master of Arts in Environmental Science from the University of South Africa. His research interests include water management, livelihoods, disaster risk management and climate change adaptation. He is a member of the Southern African Climate Change Network (SACCNET), Ecohealth Network in East and Southern Africa (ECOHESA), and the Climate Reality Leadership Corps. He is currently involved in a number of research projects with particular focus on human-environment interactions.

Barbara Ntombi Ngwenya, PhD

Dr Barbara Ntombi Ngwenya is research Professor in Applied Anthropology at the Okavango Research Institute, University of Botswana. She has a PhD in Social Work and Social Science (Cultural Anthropology) from the University of Michigan, Ann Arbor, and Masters in Social Work from Dalhousie University, Nova Scotia Canada and BA Humanities (University of Botswana). Professor Ngwenya's research interests are in socio-economic and environmental determinants of health, indigenous knowledge systems and human health in the context of vulnerability to climate change and gender and natural resources management. Professor Ngwenya is currently involved in research on indigenous knowledge systems policy and communicable diseases.

Ataur Rahman, PhD

Dr Ataur Rahman is an Associate Professor in Water and Environmental Engineering at the University of Western Sydney, Australia. He has over 20 years' experience in water industries research. He obtained his PhD degree in Hydrology from Monash University in 1997. His research interests include flood hydrology, urban hydrology and environmental risk assessment. He has over 230 research papers, reports and book chapters in water and environmental engineering. He is involved in the preparation of the new edition of the Australian Rainfall and Runoff.

Shang-Shu Shih, PhD

Dr Shang-Shu Shih is an Associate Research Fellow at the Hydrotech Research Institute at National Taiwan University (NTU); Adjunct Assistant Professor, Department of Civil Engineering at National Ilan University (NIU). He obtained his PhD from the Department of Civil Engineering, NTU in Taiwan. He joined the Department of Landscape Architecture and Environmental Planning, University of California, Berkeley, USA, as a visiting scholar during his postdoctoral research. His research interests include hydrodynamic and sediment transport simulation, eco-hydraulics, eco-hydrology, wetlands hydraulic efficiency, wetlands and river restoration. He has authored or co-authored over 20 peer-reviewed journal articles in his areas of expertise.

Olekae Tsompi Thakadu, PhD

Dr Olekae Tsompi Thakadu is Senior Research Scholar at the Okavango Research Institute, University of Botswana. He has MSc in Environment and Development from the University of Natal and PhD (Environmental Communication) from the University of Florida. His research interests are in environmental communication and education, risk management information, community-based natural resources management and indigenous knowledge systems.

EDITORS' CONTACT INFORMATION

Dr. Moseki R. Motsholapheko
Research Scholar Water Resources Management
Okavango Research Institute (ORI)
University of Botswana
P/Bag 285 Maun, Botswana
Tel: +2676861833
Fax: +267 686 1835
Email: rmoseki@ori.ub.bw

Dr. Donald Letsholo Kgathi
Professor
Okavango Research Institute
University of Botswana
P/Bag 285, Maun, Botswana
Email: kgathi@mopipi.ub.bw; dlkgathi@orc.ub.bw

INDEX